THE PHANTOM BRAKEMAN

And Other Old Time Railroad Stories

Grand Union Station, St. Louis, 1894.

EDITED, WITH INTRODUCTIONS
BY MICHAEL GILLESPIE

Published by
Great River Publishing
W987 Cedar Valley Road
Stoddard, WI 54658
608-457-2734

On the Internet, please visit www.greatriver.com

ISBN 978-0-9711602-6-2

First Edition
Designed by Sue Knopf, Graffolio, La Crosse, Wisconsin

Printed in the United States of America

Permissions

Photos courtesy Library of Congress unless otherwise noted.

Dedicated to the memory of my father,
Leo M. Gillespie
(1918-1999)

He was, in turn, a farmer, a pilot, and a steelworker.
*But he always **wanted** to be a railroad engineer.*

Other books by Michael Gillespie published by Great River Publishing, Stoddard, Wisconsin

Come Hell or High Water

A Lively History of Steamboating on the Mississippi and Ohio Rivers

ISBN 978-0-9620823-2-0

Wild River, Wooden Boats

True Stories of Steamboating and the Missouri River

ISBN 10: 0-9620823-7-6
ISBN 13: 978-0-9620823-7-5

Old Time Railroad Stories

An Anthology of True Adventures, Humorous Tales, and High Melodrama Written by Those Who Lived the Era

ISBN 978-0-9711602-5-5

Dead Under His Cab

And Other Old Time Railroad Stories

ISBN 978-0-9711602-7-9

CONTENTS

Preface

There was a time when nothing moved faster than a train. It was the era of the railroad, which, give or take a decade, lasted from the 1870s through the 1910s. In those faded times the arrival of the daily passenger train was the high point of most any day in small-town America. It was a time, too, when just about every farm boy within earshot of a fast moving freight yearned to escape the plow and ride to fame and glory as an engineer or brakeman.

Nor was this fascination limited to a rustic few. Fledgling authors found it a field ripe with possibilities. Here was action and adventure waiting to be written—tragedy, melodrama, comic relief—all wrapped up in the swirling, swaying rush of a passing train. The pulp market of the day eagerly published prose or pondering—fact or fiction—about the railroad. And even the literary magazines deemed the subject to be worthy of occasional attention.

The period in which these tales were written was a seminal time for the railroad industry. The late-nineteenth, early-twentieth century period marked a significant era of transition. The era began with a strong belief that no form of transportation could possibly replace the railroad—it was thought to be the ultimate means of moving people and goods, and would grow more or less continually. In contrast, waterborne transportation was limited to a relatively small corridor and was subject, in all but the lower latitudes, to seasonal limitations. Land transportation, via the existing roadway network, also was subject to the whims of climate. Too, there was an inherent limitation to roadway transportation, for a team of draft animals could pull only a fraction of the tonnage assigned to a typical railroad train. And what else was there? Steam tractors? Airships? These hardly seemed practical. Only the automobile could challenge the railroad industry. But the coming age of the automobile, though poised to make a monumental change in the American way of moving people and things, had not yet fully developed when the last of our stories was penned.

The era here encountered also saw a great deal of transition in the purely technical aspects of railroading. In the 1890s Congress passed the Railroad Safety Appliance Act. This mandated the use of automatic couplers, as opposed to the simple yet dangerous link-and-pin system. The link-and-pin probably contributed to more injuries and deaths on the railroad than any other device in common use. The act also required the installation of a braking system whereby the engineer could stop his train from the cab. This, in essence, meant equipping most cars with air brakes. Another part of the law called for grab irons to be placed on the ends and sides of the cars. These contrivances, as obviously necessary as they seem today, were by no means taken for granted at the dawn of the progressive era.

And yet some seemingly understandable changes did not make their way onto the railroad scene as quickly as one might expect. For instance, electric lights and steam heating on passenger cars were far from the norm when most of our tales were written. Dynamos for generating electricity throughout a train generally did not make the scene until after World War I; until then even the engine headlight was nothing more than a large lantern with a mirror reflector. Wood or coal burning stoves heated most passenger equipment, and were quite prone to set the cars on fire if they overturned in a wreck.

The hours of service laws did not come into being until 1907, and even then a sleep deprived railroad crewman could be compelled to work up to 16 hours in any 24-hour period. Prior to that, nothing could prevent a road from forcing an employee to work around the clock.

Unions, representing the various railroad crafts and trades, were still struggling to gain acceptance during the whole period. The railroad companies deemed unions to be anti-Capitalist, even subversive. Any railroad employee who joined a union did so at his own risk, for to go on strike was paramount to quitting one's job. Most rail union employees would pay their dues and hope, through some favorable legislation, to attain settlement for the many unaddressed grievances they held against their employers. Railroad union journals were the principal means of connection for this far-flung membership. Regular features in these publications included shop talk, legislative advances, lodge news, travelogues, notes of interest for wives, and some of the very stories found herein.

What follows is a gleaning of the school of railroad literature published during the height of the era. The stories are taken not only from the journals and magazines of the trade, but from the popular press of the day. All of these stories have some basis in fact, and in that regard they may be considered historically compelling.

The stories in this anthology, even the ones written as newspaper reports, were meant as entertainment, first and foremost. Whether they could be taken seriously or not mattered not so much as the fact that they amused the reader. That was their contemporary purpose, and no apologies for digressions from fact were expected, nor given. But across the chasm of time their purpose has changed. They are now as much a chronicle of their day as once they were an amusement—a folk history, if you will—coarse and incomplete, but tantalizing. For what they often lack in formality, they make up in human detail. They open a window to a pattern of life that is no more.

Some of the accounts have been edited. This was done either for considerations of length or clarity, especially when a more concise wording would better conform to the mind of the present-day reader. In either case, the edits were made with due consideration to the original intent of the author.

These old time railroad stories are tales of righteousness and morality set amidst the background of coal smoke and the incessant clatter of the telegraph sounder. Some may regard these narratives as musty remnants of a simpler time, but as the reader will see, they are full of life and emotion, and speak to each generation in ways that never change.

Reminiscences of a Pullman Conductor

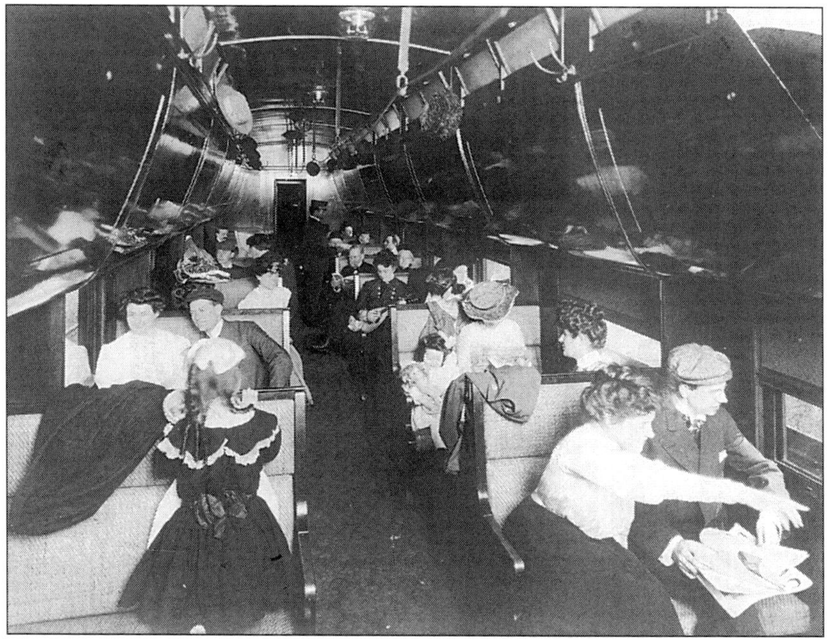

The "different types of humanity" on a Pullman sleeper, 1905. At night the upper berths will fold down from above; the facing seats will be made into lower berths. Curtains and partitions will divide the sections.

By the turn of the century, most railroads had handed over a good deal of their long-haul, overnight passenger service to the Pullman Company. The Pullman Company not only built their own cars, but operated them as a premium service, which included a conductor for the trip (in addition to the home railroad's own conductor) and a porter for each car.

Needless to say, the Pullman conductor and porters had to be adept at handling difficult passengers and unusual situations. And that delicate skill did not come easily.

Herbert Holderness, a young man seeking to make his way in life, discovered these things the hard way—through trial and error. He set out to become a Pullman conductor and stayed with it long enough to learn the pitfalls of the trade, after which, in 1901, he compiled his experiences into a book.

This, then, may be considered either his tribute to the craft, or fair warning to those foolish enough to have wanted to try it—

The problem of how to get through life in the easiest possible manner had for years been with me a fascinating study.

In spite, however, of the most arduous researches of an almost microscopic character, I was fain at last to come to the conclusion that like the quest after the Philosopher's Stone and Squaring the Circle, the problem was no nearer solution than when I first started. Everything had been tried. Tinker, tailor, soldier, sailor, apothecary, ploughboy; there remained but Gentleman and Thief.

The latter profession conscientious scruples prevented me from adopting, so I thought I would try that of gentleman.

But how to be a gentleman without means? That was the question.

One day at a railway station in New York I noticed a statuesque being in uniform, who might have been mistaken for a Greek god but for his trousers, directing with a majestic wave of the hand some people into a sleeping car.

(At that time I had never been in one in my life.)

His bearing was so dignified and he had so little to do, in fact practically nothing except to stand by the car door and look important, that I felt I had accidentally stumbled upon a heaven born occupation.

Who and what was this man? I inquired and was told he was a Pullman conductor.

I also noticed that he was attended by a colored gentleman (also in uniform), who appeared to be doing all the work; assisting passengers into the car with their hand baggage, etc. This latter personage I was informed was the porter.

When I looked again and saw all the crystallized importance of which that conductor appeared to be the almighty repository, I felt sad at heart.

Here was an ideal occupation! The very aristocracy of labor!

How was it possible for me, alone and unaided, to procure such a sinecure?

I plucked up, however, sufficient courage to come down to the depot again that evening and, cautiously approaching one of these Pullman gods, I inquired how an appointment could be obtained. He simply withered me with an icy stare and told me it was not to be thought of; that every other man in life wanted to be a conductor, but though millions applied only one in a million ever succeeded. In short I came away convinced that the post of Ambassador to St. James, or even the Presidency of the United States, were inferior jobs to this one, and more easily accessible.

But though downcast at the desolate prospect, I did not lose all hope, and after consultation with a friend who was on the New York Central railroad I thought I saw light breaking through the darkness. In other words, I received an introduction to a prominent railroad official who promised to "fix" me all right with the Pullman Company.

From this time on until I was actually appointed I dreamed of nothing but sleeping cars.

It may be worth while as an apt illustration of the illusive nature of daydreams to record the actual occurrences which took place during the initial period of my entrance into the service. Instead of bursting upon the public gaze as a full fledged conductor invested with all the powers of a relentless autocrat, I identified myself on the first night I was in charge of a sleeping car as merely a rattled mass of nerves in human form, holding in my hand a ticket punch and dimly conscious of a helpless, almost hopeless dependence on the car porter who, thank heaven, happened to be good-natured, though he gave ample evidence of his ability to see the comic side of the situation.

How I got through the horrors of that night will forever remain to me a profound mystery. Railroad tickets which should have been marked with the number of berth occupied by each passenger were taken up regardless of location or destination, and after an hour of seemingly endless work during which time the perspiration literally threatened in its profusion to leave nothing of me except the proverbial grease spot, I found myself standing dazed and stupefied with a fist full of all kinds of tickets (railroad and Pullman) inextricably tangled together and with little hope of salvation from any quarter.

To make matters worse the train conductor, who was of that stern uncompromising variety known as "Pennsylvania," treated me very brusquely and seeing the mess things were in harshly demanded what I was going to do about it? He didn't know, he was sure, for his part, why the Pullman Company could not employ someone with more sense, etc.

Right here was where all the dignity and importance I had pictured flew away, and I felt that a collision at that moment would have been an enviable distraction.

It would be futile to describe the depths of misery to which I was transported that memorable night and as morning dawned I sat in the deserted smoking room sulkily nursing my wrath and listening to the music of the snores emitted by my faithful porter, who lay asleep on the lounge and whose excessive somnolence was no doubt induced by the fatigue he had experienced by the recital of my woes and the endeavors, happily successful, to set matters right in my behalf. Poor man! Though I was supposed to report him for sleeping on duty my gratitude was such that I let him sleep on. Had I continued to feel as I felt then he might have been sleeping yet.

It is a long lane which has no turning, and after two or three trips I began to gain confidence and gradually emerged from the chrysalis state. By this time, however, I found that all my preconceived ideas were shattered and that instead of being myself important I was less even than the smallest passenger.

Publicity photo showing a "tourist conductor" with his charges, 1909. The train's real conductor is in the seat behind them.

I also at this time became convinced of the paramount importance of civility and politeness to passengers regardless of their attitude towards myself, and I may here remark that to this discovery I owe all that has been pleasant in my relation with them.

Another discovery destroyed the fallacy that in order to be a proper conductor it was necessary to bully and soundly berate the porter in the presence of passengers. This mode of procedure I early learnt was a mistake liable to end in a black eye, a broken head, or other little pleasantries at the porter's hands—a denouement not likely to lead to success with either the passengers or the company.

Here, by the way, it might be pertinent to remark that passengers upon entering a sleeping car usually make the very natural mistake of supposing that the conductor is the master hand who guides and controls all within its precincts.

In one way this is true, but on the other hand, he (the passenger) has not counted on the porter. That functionary is in reality the *deus ex machina*, whose word is law and by whose frown or favor the passenger is either very comfortable or supremely unhappy.

But to resume. After becoming somewhat proficient and having successfully strangled the various foolish ideas with which I was overcome before I entered the service, I began to try and analyze, if possible, my surroundings and to ascertain the bearings of the rocks and shoals, which I was assured on good authority, abounded on every side. (It was not long before I ran aground on one in the shape of a lost railroad ticket and it cost me $10 to get off again.)

For the benefit of conductors and in the hope that they may keep safely afloat, I append the following chart:

1. Sleeping Car Inspectors. These are derelicts which may be encountered in spite of a good lookout at any hour of the day or night.

2. Passengers who don't know what they want and insist upon getting it.

3. Chronic kickers who always have a grievance and religiously report conductors upon general principles.

4. Letting passengers off at one station when they should have got off at another. (This is a dangerous rock and often expensive.)

5. Losing railroad tickets (another dangerous rock).

6. Failing to please a car full of passengers with a temperature exactly suited to his or her individual taste.

7. Letting dust accumulate in summer when it comes in faster than it can be kept out.

8. Letting porter snore so loudly that he wakes even himself.

9. Letting porter sleep at all.

10. Failing to give a pillow to a passenger in the daytime when you see he does not want one.

11. Letting porter shine shoes in the smoking room when it is impossible for him to find a place elsewhere.

12. Letting passengers put their clothes in unoccupied upper berths even when there is plenty of room, or wrangling with them to keep the rule which says they may not do so.

13. (Being caught) Taking a drink of whisky even when you are nearly dead with the cramps.

14. Taking too much whisky whether you have cramps or not. (This is a rock upon which the ship is often lost.)

15. Trying to keep your self-respect with unreasonable or drunken passengers.

16. Absolutely keeping your self-respect and incidentally kicking an assistant superintendent or other minor Pullman official for mistaking you for a dog in his manner of speech.

17. Being too affectionate to female passengers. (This is a very hard rock to get off when you are once aground, and in a heavy sea the ship has been known to beat itself to pieces.)

18. (Being caught) Smoking on duty.

19. Putting the company's money in the wrong pocket. (This is a dangerous quicksand where the ship often sinks out of sight.)

20. Being detected by a spotter (or special agent) in any of the offenses enumerated in the above list in addition to those his own fancy may picture.

Thus only one conductor out of every hundred brings his ship safely into port. To reach the desired haven a complete equipment combining the shrewdest intelligence with a thorough self-command and many varied accomplishments thrown in is necessary. The injunction, "Be ye wise as serpents and harmless as doves," should be nailed up among the other texts in every Pullman conductor's room.

• • •

Who that has ever looked into a kaleidoscope has failed to notice the many changes of design and color which a mere turn of the instrument reveals?

Fully as wonderful is that human kaleidoscope, a Pullman car, which every trip presents an infinite variety of character for study and contemplation. The pompous, the selfish, the purse proud, and the mean are here mingled with the jovial, the gay, and the spendthrift elements of society, often in such curious combination as to make the most startling contrast, in which life's sunshine and shadow are continually chasing each other.

Those two elderly ladies who have a corpse in the baggage car and whose eyes are still wet with tears of sorrow sit in the section opposite to some giddy college youths who, with buoyant spirits and happy anticipation of the freedom they have just begun to taste, are laughingly retelling the last new joke perpetrated on one of their companions. The jaundiced, dyspeptic-looking man who sits opposite to them frowns ominously as if he felt their hilarity a personal stricture on his on ill temper, which there is little reason to doubt it really is, while the fat, perspiring lady with the butcher's shop complexion, hard by, is looking stealthily at them sideways as if to find out whether their mirth has any relation to her superabundance of adipose tissue or not.

A little meek-looking old lady with chin whiskers, who has been deposited carefully in her seat by anxious relatives before the train started, sits in trembling apprehension of the approach of the conductor to collect the tickets, unable quite as yet to reconcile herself to her new surroundings and looking every now and then to see that she is quite certain her ticket is safe.

A pompous old gentleman, accompanied by a pair of gold spectacles and a young man, appears in the car. He has purchased a ticket for the drawing room and wants to let everybody into the secret. He is fat and plethoric in appearance and inquires of the porter in a commanding voice where the conductor is to be found. If that functionary does not at once appear, the old gentlemen fusses and fumes and lets the whole car know that he thinks it is very extraordinary that no one is on hand to take his ticket and show him his stateroom. "My stateroom; my stateroom." That is the burden of his song, as if the world should and would stand still to attend to him—the only one in it.

Who has failed to observe the gay and glittering female who languishes in solitude the first fifty miles of the journey, casting around ever and anon glances so melting and appealing as to soften the stoniest masculine heart, afterwards to be found by some ardent admirer who, like his prototype, the stage hero, has a knack of appearing at the most opportune moment and to be "quite unexpected."

"So charmed to see you, Mr. Dash! Why, I thought you were in St. Louis!"—although in all probability she had watched him get into one of the day coaches at the terminus.

Look at that precious piece of humanity sitting all by himself huddled up in the corner of his seat with a woolen comforter, although the weather is warm, tightly wound around his throat.

How uneasily he glances up every now and then at the ventilator overhead if to make sure that not a breath of air can reach him.

"Porter! I feel a draught coming from somewhere. Shut those ventilators immediately, and I protest against that open window over there."

Other people may suffocate with the heat, but it makes no difference to him. He has never taken a bath in his life and air might prove fatal to him.

Sometimes the emigrant father, the possessor of dirt and dollars, makes his appearance usually accompanied by a malodorous family and a frowsy-looking woman, bringing with them enough bundles and parcels to fill an entire baggage car. They are coming from somewhere outside the confines of civilization and are at once the bugbear and despair of the porter, who is continually cleaning up after them. As they generally speak a confounded jargon of Polish and Bohemian and do not understand any other language, it is a difficult matter to head them off in any direction or in fact to deal with them at all.

The invalid who was brought in on a stretcher at the last stopping place is in the final stage of consumption and is going home to die. The great emaciation, unusual brilliancy of eye, the hectic flush and paroxysms of coughing, followed by those telltale drops which dye the handkerchief held to his mouth, all proclaim the approach of the destroyer. He has come all the way from Denver accompanied by his devoted young wife. Watch with what assiduous attention she notes every look, and the air of cheerfulness, I had almost said gaiety, with which she waits upon him in order that he may not guess the existence of that sorrow which is gnawing at her heart.

These examples are but a few caught at random, and form a very infinitesimal portion of the different types of humanity daily to be met with on a Pullman car.

I have a very vivid recollection of having once at a terminal station assisted an old lady into the car apparently without any other baggage than a small hand satchel, afterwards, however, to discover that a tall parcel evidently with a dome-shaped top, but carefully concealed in newspaper wrappings had been deposited by someone in the section reserved for her.

Times out of number the old lady was seen to peep very cautiously into the parcel and utter a few cabalistic words, which sounded to me more like an incantation in the Ojibwa language than anything else, and every time she did so curious noises proceeded from the interior, leaving me still unenlightened and more mystified than ever.

When the berths were made up at night and the old lady was forced to move in order to have her own attended to, she hastily seized the parcel and held it on her lap until she could retire, which she did as soon as possible, placing it carefully inside the berth curtains. This action was seriously opposed by the porter who, however, was finally overcome almost to tears upon finding a dollar thrust into his hand and very soon the old lady and her mysterious companion were apparently—to judge from the silence which reigned—"off to the land of nod."

About the hour of 2 o'clock on the following morning I was sitting (for it was my watch on deck, so to speak) in an unoccupied section of the dimly lighted car soliloquizing alternately on the wonderful power of the human nose as a tone producing instrument and sinking every now and then myself into a nodding doze when I was suddenly startled by the sound of a voice singing and then stopping suddenly with a "whirr" like a policeman's rattle.

Being half asleep and half awake I thought it probable that my mind had played me false, and after attributing the sensations I experienced to a pork pie hastily swallowed at the last stopping station, I had once again composed myself for another nap when a hoarse, sepulchral voice was heard to utter very slowly and distinctly these words: "Betsy, prepare to die!"

A death-like stillness ensued, broken only by the audible thumping of my own heart coupled with a dreadful sense of oppression which twenty pork pies could never have produced.

"Murder! Murder! Save me!" gasped a voice in smothered accents. There was no mistaking the realism of the whole thing this time, and kicking over hastily all thoughts of indigestible pastry, I stood bolt upright shaking all over with apprehension and large beads of perspiration.

Here was a dreadful situation!!! There appeared to be no evidence of a struggle going on in any of the berths, and rushing frantically up and down the car I at length arrived trembling with excitement at the spot from whence I thought the voice had proceeded, but could find nothing.

By this time the passengers were all fairly aroused and just as I was about to institute a thorough search throughout each berth, a voice in the old lady's section was heard singing, "Over the garden wall I let the baby fall."

It was suggested by some that the occupant might be an escaped lunatic, but upon investigation it turned out that the mysterious parcel which accompanied the old lady was nothing more than a cage with a parrot in it, who, unused to being cooped up so long in such a confined atmosphere, had tried to relieve his feelings by giving a series of recitations.

The old lady on being aroused was heard to soundly berate the unfortunate bird, who rejoiced in the name of "Buller," and threatening all kinds of dire vengeance against the company and its officials, retired with her companion to the baggage car to await, as best she could, the morning light.

How it happened that I got out of that scrape with either my reason or my position can never be satisfactorily explained, except on the theory that the exceeding comicality of the situation proved my salvation.

Hereafter all mysterious parcels underwent a rigid inspection to the utter exclusion of pet animals in general and parrots in particular.

Herbert O. Holderness, *The Reminiscences of a Pullman Conductor, or Character Sketches of Life in a Pullman Car* (Chicago: n.p., 1901), pp. 9-26, 38-42.

The Parlor Car Romance

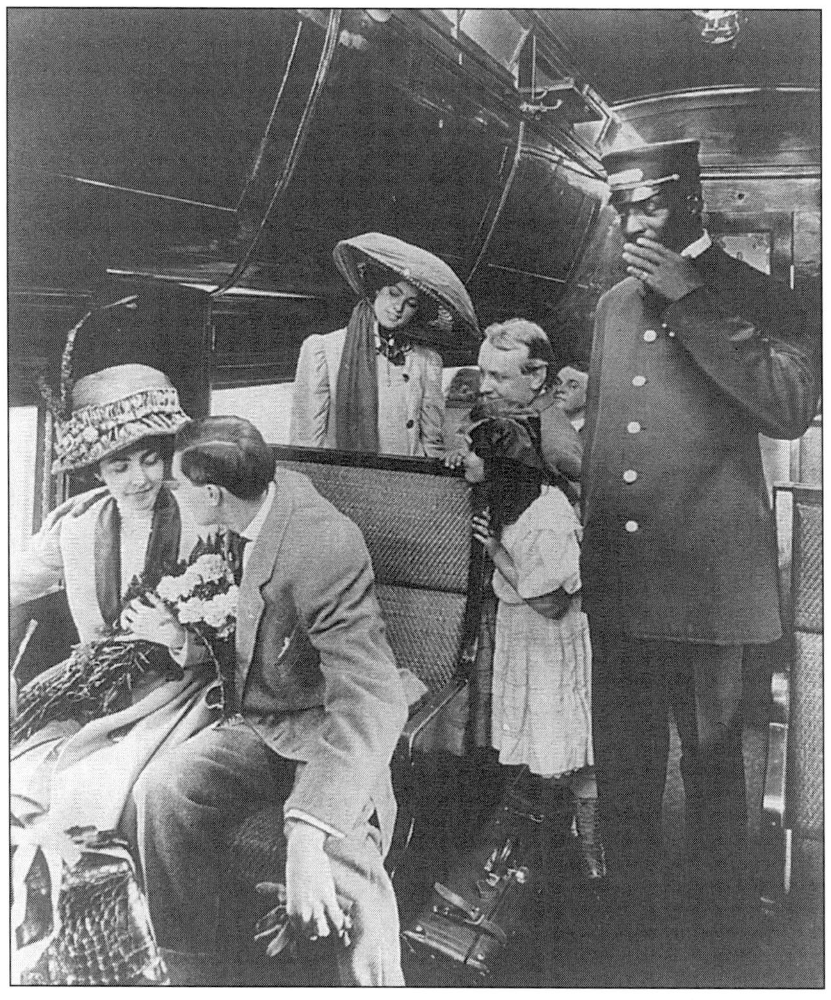

Honeymoon bound on a sleeper, 1909.

There is more than one way to interpret the phrase "romance of the rails." While it is most often a figurative term, there comes perchance the occasional literal application. And that is what we now encounter. This is not the story of a train, but a story that takes place on a train. It involves a gentleman suitor named Sutton, and Betty, the object of his attentions. And, too, there is Betty's young brother, Tommy, who has a way of spoiling all of Sutton's advances toward the coquettish Betty. Try as he may, the dashing Sutton can't seem to get his point across, in...

Tommy loved "Big Sister," and so did Jack Sutton. Betty Travers loved Tommy openly and Jack Sutton secretly. Tommy liked Sutton, but somehow this affection was not reciprocated. And this was the way matters stood when Sutton came upon Betty and Tommy in the Chicago sleeper.

"Such a pleasant surprise," smiled Betty, as she extended a smoothly gloved hand in greeting. "I was afraid it would be such a dreary trip."

Sutton glowed.

"We can have lots of fun," supplemented Tommy, slipping a moist and uncomfortably sticky hand into Sutton's.

Sutton glowered.

Betty Travers was not the sort of girl to wear her heart upon her sleeve. For more than a year Sutton had sought to learn his fate, and Betty, woman-like, had parried his every attempt. This was his opportunity. If only Tommy—

"Won't you sit down?" asked Betty, hospitably making room for him on the seat across the way.

"I'm sorry," said Sutton, boldly, "but I cannot ride with my back to the engine."

"I can ride with my back to the engine," said Tommy, proud that he possessed accomplishments not enjoyed by his elders. To prove it, he slipped into the facing seat, and Sutton sank down beside the girl. He felt a sudden desire to buy Tommy candy, but the newsboy was on the car ahead.

The rush of the train made it possible for Sutton to speak freely, and after a few conventional remarks he plunged boldly into the subject next his heart.

"Do you know," he asked, "that I have wanted just this chance to talk with you alone for a long time?"

"Solitude in a crowd," she suggested uneasily, glancing at a man across the aisle who seemed unduly interested.

"This is better than a ballroom," he protested, "or at theater. I want to ask you if—"

"Look at the cows!" shouted Tommy, precipitating himself into his sister's lap, and pointing wildly at the flying landscape.

"I wanted to ask you," continued Sutton, "if there is anything I have done to cause you to avoid me, you—"

"I've got a cow, too," shouted Tommy from the opposite seat. "I keep it out at grandma's."

"That's a good place for it," said Sutton to Tommy. Then turning to Betty, he resumed: "You seem to be afraid to give me a moment alone with you. Do you fear that I shall—"

"Do tricks," demanded Tommy, climbing upon Sutton's knee. Sutton performed some simple parlor magic, and the boy, satisfied, turned his attention to the window.

"Are you afraid of the question I want to ask, and which means so much to me?" he urged.

"I have not tried to avoid you," she said, blushing prettily. "I'm sure I've given you a lot of my time this season."

"Yes," he agreed, "but for months I have tried to tell you something, and you have evaded me just as you are trying to do now."

"I don't think this is evading you," she laughed, glancing down at the broad shoulder so close to her own.

"But this is the first time," he persisted, "and you couldn't help yourself. You must have seen that—"

"We're going over a big river!" shouted Tommy, gleefully. "Look at the boats."

Betty took advantage of the diversion, and it was several minutes before Sutton could lead the talk back into the proper channel.

"We've been friends for so long," he said finally. "Won't you tell me whether there is any possibility of my ever—"

"Going to stop," announced Tommy, as the train slowed down for a station. "Take me for a walk."

"Please do," she pleaded. "The poor little fellow gets so tired and restless."

Dutifully Sutton led the lad up and down the platform until he saw the conductor coming away from the telegraph window. Ten minutes more were wasted by the young torment in recounting his experience to Betty.

At last he subsided, and once more Sutton sought to learn his fate. "Don't you know," he pleaded, "that I have loved you ever since I've known you? Can't you say one little word to give me hope that some day—"

"Lemme your watch," demanded Tommy. Sutton passed over the handsome timepiece without a word. If it would keep the boy busy, he did not care what happened.

"—That some day," he went on, "I may win you for my wife? I have thought sometimes that I have been so fortunate as to find favor in your eyes, and then again I have—"

"Wheels!" demanded Tommy, handing back the watch. With a sudden memory of what the back case contained, Sutton explained to the boy that the case did not open in the back. Betty's face clouded. It might hurt the works to open the case in all this dust, but still—

Tommy had discovered the way to make the hands go around by depressing the stem, and he was happy again.

"There have been times," continued Sutton, "when I have feared that your avoidance of me was intentional, and that you wanted to save me the pain of refusal. No matter what your answer may be, I want to know my fate; to know whether I have any chance to win the sweetest woman in the world. Won't you tell me that you do care for me—that you will—"

"Wind it up," commanded Tommy. With visions of dreadful fates, any one of which was too merciful for the boy, Sutton took the watch and held it to his ear. The works had stopped.

In desperation he showed the boy how to depress a boss on the side, and make the repeater strike the hour and minute, and Tommy, beaming, went back to the other seat.

"Won't you give me an answer?" Sutton pleaded. Her eyes grew misty. She wanted to answer, but somehow she dreaded telling him on the car. The same feeling that had led her to parry his proposals now impelled her to temporize. It was Tommy who solved the situation. The watch flew from his hands and clattered to the floor.

"It just jumped out of my hands," explained the frightened boy, as he slid from the seat to pick it up. "The back's broken off," he added, in dismay.

Sutton sprang for it, but the boy was too quick for him. "It's got a picture, Betty's picture," he cried as he turned the cap over. "Do all watches have pictures in them?"

"No," said Sutton, shortly, snapping the cap back on the case.

"Betty's has," persisted the boy.

"Tommy, you mustn't fib," she cried, turning a rosy red.

"Yes, it has," he insisted, snatching at her chatelaine. "See!" And he tore the back case open.

Sutton gave one look at the picture of himself, and then one longer look into Betty's tender eyes.

"God bless you, Tommy," he said, very gently.

Epes W. Sargeant, "A Parlor Car Romance," *The Railway Conductor* 33, no. 8 (August 1916): 558-60.

When Doctors Agreed

In the heyday of passenger railroading, few things were as sought after as a railroad pass. The rail companies were surprisingly generous with their passes, but often with a hidden motive. For example, a railroad might present a pass to a physician with the expectation that the good doctor would not fuss at being aroused in the middle of a journey to look after a sick passenger. Yet the chances of that happening were rather slim, were they not? So who would be the wiser if the free-riding physician was... other than a physician?

Mr. Tecumseh Clay had never traveled on a railroad pass, though he had often wished that he might. So when Dr. Erasmus Evans, who had an annual pass on the AB&C road, offered to let Mr. Clay use it, the offer was eagerly accepted.

"The pass is non-transferable," said Dr. Evans, "but that won't make any difference. Just pretend you are me if the conductor says anything; but he won't."

Mr. Clay took the night train, due in St. Louis the next morning. He awaited the advent of the train conductor in some trepidation, wondering to what extent he might have to prevaricate should the official prove to be of the extra-inquisitive type. Mr. Clay didn't like to lie, and hoped the conductor wouldn't make him. At the same time he was a determined man, and did not intend that a fib or two should stand in the way of a free ride. Besides, the safety of the doctor's pass might be imperiled if he exhibited any weakness or confusion during the possible cross-examination.

But when the conductor appeared he merely read the name on the proffered pass, returned it to Mr. Clay, and went on, leaving Mr. Clay rejoicing. Not even the littlest and snowiest of fibs had he had to utter. So Mr. Clay, with a pleasant consciousness of both thrift and rectitude, settled comfortably back on the cushions in his section of the sleeper, and presently, having let the chocolate-faced porter make up his berth, he crawled into such slumber as the rushing train might permit.

About midnight he was aroused by a voice at the curtains of his berth. "Doctor!" it said. "Doctor! Wake up! A man in the next car has been taken sick, and needs something done."

It was the conductor, who had noticed that the name on the pass carried an M.D.

"All right. I'll be out in a moment," answered Mr. Clay, with a promptitude that surprised even himself. "The dickens!" he muttered, when the conductor had departed. "Why didn't Evans tell me that doctors are called up in the middle of the night on sleeping cars just the same as anywhere else? I'd have let him keep his pass and paid my fare if I'd known. There's nothing to do, though, but go and see the man. If he's really sick enough to need a doctor I'm sorry for him."

Mr. Clay, having dressed hastily, made his way into the next ear, and was conducted to the patient. With commendable gravity he felt of the man's pulse, placed his hand on his chest, and counted the respirations, and then asked to see his tongue. This done, he stood for a moment gazing contemplatively upon the luckless patient. The bystanders thought he was pondering deeply; he was really wondering what he should do next. Then—it came like an inspiration; he had seen Dr. Evans do it one time—he lifted the patient's hand and studied his fingernails in a meditative manner.

"Have you some whiskey?" he asked, turning to the conductor.

"Yes, sir; I can get some," was the answer.

"Very good! Give him two teaspoonfuls in half a glass of water, and repeat the dose at the end of an hour. I haven't my medicine case with me, unfortunately, and can't prescribe just as I'd like to. But the whiskey will act as a—"

What sort of an actor the whiskey would prove he evidently regarded as of no great importance to his listeners, for he broke off, and remarked that he was sorry he hadn't his thermometer with him; he would like to take the patient's temperature. He evidently had some fever. "But give him the whiskey as directed," he concluded, with brisk decisiveness, "and if there should be a change for the worse, let me know."

Back in the privacy of his berth once more Mr. Clay smiled broadly, and then sighed deeply. "Poor fellow," he thought. "I hope it's nothing serious."

"Doctor!" called a voice, just as he was dozing off. "The man seems to be getting worse. I guess you'd better take another look at him."

"All right," answered Mr. Clay, cheerfully, but groaning inwardly. "I wish," he muttered, "that confounded old pass had been taken up and cancelled before it ever fell into my hands! What the deuce am I to do, anyway? The man may die for lack of a little medical skill. But I can't confess that I'm no doctor; I've got to bluff it out."

"There's another doctor in the forward car, sir," said the conductor as Mr. Clay appeared. "The patient's friends are getting kind o' nervous, and thought perhaps you'd like to consult with him. I'll rout him out if you think best."

"Very well, if the patient's friends desire it," answered Mr. Clay, both relieved and annoyed. "That doctor will see through me in about thirty seconds," he reflected, gloomily. "I wonder if it would kill a man to jump off the train; it's going pretty fast."

But Mr. Clay did nothing so rash as that. He was gazing calmly at the patient when the consulting doctor arrived. "This is Dr. Evans, Dr. Brown," said the conductor, guiltless of intentional falsehood.

The two professional men bowed gravely to each other. Dr. Brown had brought a small medicine case with him, which he set down in the aisle. "Well, Dr. Evans, what are the symptoms?" he asked.

"Just take a look at him and see what you think, Dr. Brown," replied Mr. Clay, with admirable self-possession.

Dr. Brown drew a fever thermometer from his pocket, shook the fluid down with a quick professional jerk, and inserted the end under the patient's tongue. Then he felt his pulse, and Mr. Clay noted with envy that he did not look at his watch, as he himself had done. Mr. Clay recalled that Dr. Evans seldom looked at his watch while counting a patient's pulse.

"What has been done for the relief of the patient, Dr. Evans?" asked the consulting physician, as he withdrew the thermometer and silently studied the temperature registered.

Mr. Clay told him. Doctors had disagreed before, and they might as well do so again, reflected the unhappy Clay. Besides, there was nothing else to do but tell him.

Dr. Brown made no comment for a moment. He seemed to be considering the case carefully. Presently, to Mr. Clay's relief and astonishment, he said: "Well, I think you did the right thing. I should advise continuing the treatment through the night, and if the patient hasn't improved by morning we can decide upon further treatment. His temperature is not alarming."

So back to his berth, conscious that Providence was kinder than he deserved, went Mr. Clay. If his views as to the patient's condition were hazy, upon one subject he held a definite opinion; he was determined never again to travel upon a physician's pass.

The next morning the patient was reported very much better, and Mr. Clay's heart overflowed with gratitude. As he left the train he met Dr. Brown. They passed through the station together, and as they started to part on the street, Mr. Clay said, with a confidential smile:

"Between you and me, Doctor, I'm not a physician at all. I couldn't tell the conductor so, though, because I'm traveling on a physician's pass."

Dr. Brown's lips twitched, and he held out a cordial hand. "I brought along this medicine case," he said, "just as a bit of a bluff. I'm no more of a physician than you are, but I'm traveling on Dr. Brown's pass!"

James R. Perry, "When Doctors Agreed," *Harper's Monthly Magazine* 104, no. 620 (January 1902): 348-49.

The Invalid's Story

An express baggage car in its usual position—the first car behind the engine and tender. Rock Island Railroad, 1909.

It was a seldom advertised, but highly practical fact, that passenger railroads, right up into the days of Amtrak, carried human remains as a matter of course in their baggage/express cars.

This service, which was performed for the usual express charge, rarely drew the attention of railroad writers. But Mark Twain was no railroad writer. And if there was any possibility of humor in the deed, he would certainly flush it out. Never mind that distinguished Eastern critics found the following story distasteful; most readers couldn't restrain their laughter. And so, for those whose sense of decorum is not easily offended, we present this 1880s tale of mixed-up freight—

I seem sixty and married, but these effects are due to my condition and sufferings, for I am a bachelor, and only forty-one. It will be hard for you to believe that I, who am now but a shadow, was a hale, hearty man two short years ago—a man of iron, a very athlete! Yet such is the simple truth. But stranger still than this fact is the way in which I lost my health. I lost it through helping to take care of a box of guns on a two-hundred-mile railway journey one winter's night. It is the actual truth, and I will tell you about it.

I belong in Cleveland, Ohio. One winter's night, two years ago, I reached home just after dark, in a driving snowstorm, and the first thing I heard when I entered the house was that my dearest boyhood friend and schoolmate, John B. Hackett, had died the day before, and that his last utterance had been a desire that I would take his remains home to his poor old father and mother in Wisconsin. I was greatly shocked and grieved, but there was no time to waste in emotions; I must start at once. I took the card, marked "Deacon Levi Hackett, Bethlehem, Wisconsin," and hurried off through the whistling storm to the railway station. Arrived there I found the long white-pine box which had been described to me; I fastened the card to it with some tacks, saw it put safely aboard the express car, and then ran into the eating-room to provide myself with a sandwich and some cigars.

When I returned, presently, there was my coffin-box back again, apparently, and a young fellow examining around it, with a card in his hand, and some tacks and a hammer! I was astonished and puzzled. He began to nail on his card, and I rushed out to the express car, in a good deal of a state of mind, to ask for an explanation. But no—there was my box, all right, in the express car; it hadn't been disturbed. [The fact is that without my suspecting it a prodigious mistake had been made. I was carrying off a box of *guns* which that young fellow had come to the station to ship to a rifle company in

Peoria, Illinois, and *he* had got my corpse!] Just then the conductor sung out "All aboard," and I jumped into the express car and got a comfortable seat on a bale of buckets.

The expressman was there, hard at work—a plain man of fifty, with a simple, honest, good-natured face, and a breezy, practical heartiness in his general style. As the train moved off a stranger skipped into the car and set a package of peculiarly mature and capable Limburger cheese on one end of my coffin-box—I mean my box of guns. That is to say, I know now that it was Limburger cheese, but at that time I never had heard of the article in my life, and of course was wholly ignorant of its character. Well, we sped through the wild night, the bitter storm raged on, a cheerless misery stole over me, my heart went down, down, down!

The old expressman made a brisk remark or two about the tempest and the arctic weather, slammed his sliding doors to, and bolted them, closed his window down tight, and then went bustling around, here and there and yonder, setting things to rights, and all the time contentedly humming "Sweet By and By," in a low tone, and flatting a good deal. Presently I began to detect a most evil and searching odor stealing about on the frozen air. This depressed my spirits still more, because of course I attributed it to my poor departed friend. There was something infinitely saddening about his calling himself to my remembrance in this dumb pathetic way, so it was hard to keep the tears back. Moreover, it distressed me on account of the old expressman, who, I was afraid, might notice it. However, he went humming tranquilly on, and gave no sign; and for this I was grateful. Grateful yes, but still uneasy; and soon I began to feel more and more uneasy every minute, for every minute that went by that odor thickened up the more, and got to be more and more gamey and hard to stand. Presently, having got things arranged to his satisfaction, the expressman got some wood and made up a tremendous fire in his stove. This distressed me more than I can tell, for I could not but feel that it was a mistake. I was sure that the effect would be deleterious upon my poor departed friend.

Thompson—the expressman's name was Thompson, as I found out in the course of the night—now went poking around his car, stopping up whatever stray cracks he could find, remarking that it didn't make any difference what kind of a night it was outside, he calculated to make us comfortable, anyway. I said nothing, but I believed he was not choosing the right way. Meantime he was humming to himself just as before; and meantime, too, the stove was getting hotter and hotter, and the place closer and closer. I felt myself growing pale and qualmish, but grieved in silence and said nothing. Soon I noticed that the "Sweet By and By" was gradually fading out; next it ceased altogether, and there was an ominous stillness. After a few moments Thompson said—

"Pfew! I reckon it ain't no cinnamon't I've loaded up thish-yer stove with!"

He gasped once or twice, then moved toward the cof—gun-box, stood over that Limburger cheese part of a moment, then came back and sat down near me, looking a good deal impressed. After a contemplative pause, he said, indicating the box with a gesture—

"Friend of yourn?"

"Yes," I said with a sigh.

"He's pretty ripe, *ain't* he!"

Nothing further was said for perhaps a couple of minutes, each being busy with his own thoughts; then Thompson said, in a low, awed voice—

"Sometimes it's uncertain whether they're really gone or not—seem gone, you know—body warm, joints limber—and so, although you *think* they're gone, you don't really know. I've had cases in my car. It's perfectly awful, becuz *you* don't know what minute they'll rise right up and look at you!" Then, after a pause, and slightly lifting his elbow toward the box—"But *he* ain't in no trance! No, sir, I go bail for *him*!"

We sat some time, in meditative silence, listening to the wind and the roar of the train; then Thompson said, with a good deal of feeling—

"Well-a-well, we've all got to go, they ain't no getting around it. Man that is born of woman is of few days and far between, as Scriptur' says. Yes, you look at it any way you want to, it's awful solemn and cur'us; they ain't *nobody* can get around it; *all's* got to go—just *everybody*, as you may say. One day you're hearty and strong"—here he scrambled to his feet and broke a pane and stretched his nose out at it a moment or two, then sat down again while I struggled up and thrust my nose out at the same place, and this we kept on doing every now and then—"and next day he's cut down like the grass, and the places which knowed him then knows him no more forever, as Scriptur' says. Yes-'ndeedy, it's awful solemn and cur'us; but we've all got to go, one time or another, they ain't no getting around it."

There was another long pause; then—

"What did he die of?"

I said I didn't know.

"How long has he ben dead?"

It seemed judicious to enlarge the facts to fit the probabilities; so I said—

"Two or three days."

But it did no good; for Thompson received it with an injured look which plainly said, "Two or three *years*, you mean." Then he went right along, placidly ignoring my statement, and gave his views at considerable length upon the unwisdom of putting off burials too long. Then he lounged off toward the box, stood a moment, then came back on a sharp trot and visited the broken pane, observing—

"'T would 'a' ben a dum sight better, all around, if they'd started him along last summer."

Thompson sat down and buried his face in his red silk handkerchief, and began to slowly sway and rock his body like one who is doing his best to endure the almost unendurable. By this time the fragrance—if you may call it fragrance—was just about suffocating, as near as you can come at it. Thompson's face was turning gray; I knew mine hadn't any color left in it. By and by Thompson rested his forehead in his left hand, with his elbow on his knee, and sort of waved his red handkerchief towards the box with his other hand, and said—

"I've carried a many a one of 'em—some of 'em considerable overdue, too—but, lordy, he just lays over 'em all!—and does it *easy*. Cap, they was heliotrope to *him*!"

This recognition of my poor friend gratified me, in spite of the sad circumstances, because it had so much the sound of a compliment.

Pretty soon it was plain that something had got to be done. I suggested cigars. Thompson thought it was a good idea. He said—

"Likely it'll modify him some."

We puffed gingerly along for a while, and tried hard to imagine that things were improved. But it wasn't any use. Before very long, and without any consultation, both cigars were quietly dropped from our nerveless fingers at the same moment. Thompson said, with a sigh—

"No, Cap, it don't modify him worth a cent. Fact is, it makes him worse, becuz it appears to stir up his ambition. What do you reckon we better do, now?"

I was not able to suggest anything; indeed, I had to be swallowing and swallowing, all the time, and did not like to trust myself to speak. Thompson fell to maundering, in a desultory and low-spirited way, about the miserable experiences of this night; and he got to referring to my poor friend by various titles—sometimes military ones, sometimes civil ones; and I noticed that as fast as my poor friend's effectiveness grew, Thompson promoted him accordingly—gave him a bigger title. Finally he said—

"I've got an idea. Suppos'n' we buckle down to it and give the Colonel a bit of a shove towards t' other end of the car?—about ten foot, say. He wouldn't have so much influence, then, don't you reckon?"

I said it was a good scheme. So we took in a good fresh breath at the broken pane, calculating to hold it till we got through; then we went there and bent down over that deadly cheese and took a grip on the box. Thompson nodded "All ready," and then we threw ourselves forward with all our might; but Thompson slipped, and slumped down with his nose on the cheese, and his breath got loose. He gagged and gasped, and floundered up and made a break for the door, pawing the air and saying, hoarsely, "Don't bender me!—gimme the road! I'm a-dying; gimme the road!"

Out on the cold platform I sat down and held his head a while, and he revived. Presently he said—

"Do you reckon we started the Gen'rul any?"

I said no; we hadn't budged him.

"Well, then, *that* idea's up the flume. We got to think up something else. He's suited wher' he is, I reckon; and if that's the way he feels about it, and has made up his mind that he don't wish to be disturbed, you bet you he's a-going to have his own way in the business. Yes, better leave him right wher' he is, long as he wants it so; becuz he holds all the trumps, don't you know, and so it stands

Mark Twain on the lecture circuit, 1885.

to reason that the man that lays out to alter his plans for him is going to get left."

But we couldn't stay out there in that mad storm; we should have frozen to death. So we went in again and shut the door, and began to suffer once more and take turns at the break in the window. By and by, as we were starting away from a station where we had stopped a moment, Thompson pranced in cheerily, and exclaimed—

"We're all right, now! I reckon we've got the Commodore this time. I judge I've got the stuff here that'll take the tuck out of him."

It was carbolic acid. He had a carboy of it. He sprinkled it all around everywhere; in fact he drenched everything with it, rifle-box, cheese, and all. Then we sat down, feeling pretty hopeful. But it wasn't for long. You see the two perfumes began to mix, and then—well, pretty soon we made a break for the door; and out there Thompson swabbed his face with his bandanna and said in a kind of disheartened way—

"It ain't no use. We can't buck agin *him*. He just utilizes everything we put up to modify him with, and gives it his own flavor and plays it back on us. Why, Cap, don't you know, it's as much as a hundred times worse in there now than it was when he first got a-going. I never *did* see one of 'em warm up to his work so, and take such a damnation interest in it. No, sir, I never did, as long as I've ben on the road; and I've carried a many a one of 'em, as I was telling you."

We went in again, after we were frozen pretty stiff; but my, we couldn't *stay* in, now. So we just waltzed back and forth, freezing, and thawing, and stifling, by turns. In about an hour we stopped at another station; and as we left it Thompson came in with a bag, and said—

"Cap, I'm a-going to chance him once more—just this once; and if we don't fetch him this time, the thing for us to do, is to just throw up the sponge and withdraw from the canvass. That's the way *I* put it up."

He had brought a lot of chicken feathers, and dried apples, and leaf tobacco, and rags, and old shoes, and sulphur, and asafoetida, and one thing or another; and he piled them on a breadth of sheet iron in the middle of the floor, and set fire to them. When they got well started, I couldn't see, myself, how even the corpse could stand it. All that went before was just simply poetry to that smell—but mind you, the original smell stood up out of it just as sublime as ever—fact is, these other smells just seemed to give it a better hold; and my, how rich it was! I didn't make these reflections there—there wasn't time—made them on the platform. And breaking for the platform, Thompson got suffocated and fell; and before I got him dragged out, which I did by the collar, I was mighty near gone myself. When we revived, Thompson said dejectedly—

"We got to stay out here, Cap. We got to do it. They ain't no other way. The Governor wants to travel alone, and he's fixed so he can outvote us."

And presently he added—

"And don't you know, we're *pisoned*. It's *our* last trip, you can make up your mind to it. Typhoid fever is what's going to come of this. I feel it a-coming right now. Yes, sir, we're elected, just as sure as you're born."

We were taken from the platform an hour later, frozen and insensible, at the next station, and I went straight off into a virulent fever, and never knew anything again for three weeks. I found out, then, that I had spent that awful night with a harmless box of rifles and a lot of innocent cheese; but the news was too late to save *me*; imagination had done its work, and my health was permanently shattered; neither Bermuda nor any other land can ever bring it back to me. This is my last trip; I am on my way home to die.

Mark Twain, *The Stolen White Elephant, Etc.* (Boston: Osgood, 1882), pp. 94-104.

The Runaway

A safety switch and siding can be seen just above this curve on the Canadian Pacific Railroad, 1902. The siding sloped uphill in order to stop a runaway car or train.

Hoboes and trainmen were born enemies; it couldn't be otherwise. Hoboes were perceived as worthless freeloaders and troublemakers for whom the average "shack"—the brakeman—had little sympathy. Often, the shack, upon discovering a hobo on his train, would demand money for the ride, after which the hobo usually was told to get off—regardless of whether the train was moving or not. But the hobo—no innocent himself—had ways of getting even, no matter what the consequence.

One of the most vulnerable parts of a train was the air brake system, which had been universally adopted by the time of this story—1915. The air pressure was carried from car to car through a hose located next to the coupler. Each car had an air hose valve on either end. If a valve were closed, the air brakes would not work on the cars behind it. And every hobo knew that—

Had we known that the tramp was an escaped convict we would have watched his actions more closely than we did, after putting him off.

I had the pile-driver outfit in the train that night, as there were several bridges in the Wild Horse Canyon out of line and needed repairing. Bill Anderson, the foreman of the outfit, lived in one of the outfit cars just behind the driver-car. His wife was with him, doing the cooking for the bridge men. In the car also was Hazel, the year-old baby, who was the idol of all the men.

We stopped at Pines on account of a hot box on a car of lumber, which needed attention before proceeding farther.

While there, "Slats" Monroe, the head shack, found a tramp in an empty, and ordered the fellow to unload. The bum refused to do so and cursed "Slats" roundly. That was a very foolish thing to do as Monroe was a husky brakeman and he also carried his brake club. He promptly knocked the man out of the car and told him to beat it.

The tramp went off swearing vengeance on Monroe, but that worthy only hooted at him. The tramp then went back along the train to the cook-car and asked Mrs. Anderson to give him a "hand-out." The kind-hearted young woman gave him a lunch.

When we were ready to pull out, Anderson called to me to come over to the cook-car and share a midnight lunch with them.

As Mrs. Anderson was a splendid cook, I was only too glad to accept his offer, and at once made preparations to go.

Had I known that the bum had gone ahead and had turned an angle cock three cars behind the engine, I am afraid that the lunch would have received scant attention.

The brakemen took their places on top of the train ready to set the handbrakes while descending the heavy grade down Wild Horse Canyon. Williams, the engineer on the 1282, was as good a man on the mountain as we had on that division. "Red" Brown, the rear shack, was an experienced mountain man, cool-headed and fearless. It was generally conceded that I had about the best crew on the mountain.

Soon we rattled over the switches, and after looking at the air-gauge and seeing that the pressure was good, I went ahead to the cook-car. When we passed the first curve down the canyon I noticed that the speed was increasing to a dangerous degree and I went to the door to see if anything was wrong.

As I opened the door I could hear the siren whistle of the big 1282 calling for the hand-brakes. I then knew that we were running away, something that will send a chill through the bravest heart, for Death always hovers close to trains that have gone beyond human control.

I had carried my brake club with me, and at once I went out on top of the cook-car and clubbed up the brake. Then I tightened up on the brakes of the three cars of piling between the cook-car and the caboose. As I set up the caboose brake I looked at the gauge, and anyone can imagine my feelings when I saw that we only had ten pounds of air to control the train with.

As I went back over the second car from the caboose, I was startled by a loud sound of breaking wood. Looking back, I saw that the stakes had broken off the sides of the car of piling ahead of the caboose and the great thirty-foot pilings were flying off on both sides like giant toothpicks. I fully expected to see the caboose derailed at any moment.

Over ahead of the pile-driver I could see the two brakemen doubling up on the brakes. Williams was staying with the engine, I could see, for every few moments he would reverse the big consolidation to bunch the slack, which gave the brakes a new chance to hold. I could hear the bridge men in the bunk cars ahead of the driver crying out in their helplessness like wild animals in cages, for they could not get out.

I worked my way over to the cook-car and there found the cool-headed Anderson trying to quiet his wife, who was hysterical. He had wrapped up the baby in a little mattress to protect it in case of a derailment. The dishes were rolling about the car, and, as the speed increased, pots and pans were thrown off the hooks amid a great clatter, while all about was the roar of the runaway train.

I advised Anderson to be ready to jump if the cars began to pile up, as the chances, slim at the best, were better in the open than when mixed up with the cars. I went back on top of the cook-car to try to get another notch on the brake, when I heard a loud crash overhead. In the moonlight I could see that the pile-driver had broken loose from its mooring and was swinging sideways in a dangerous manner. Right then I knew what our end would be if we managed to hold the rails, a doubtful thing to do.

For just at the Sphinx Curve, five miles down the canyon, was a high bridge of the overhead style with a narrow clearance. I knew that the pile driver would most likely swing across the track and catch in the bridge, causing a general wreck. There was about one chance in a hundred that the big driver-frame might swing in line with the cars just as we readied the bridge.

On we rushed with fire flying from the grinding brakes and a dozen journal boxes, lighting up the rocky cuts like torches. In a

A steam crane pile driver on a bridge construction crew, circa 1910. The car's air hose can be seen dangling alongside the coupler. The angle cock valve is located at the top of the hose.

few moments we swung out onto the high line along the canyon's sides, and a mile down I could see the Sphinx Head standing out clear in the moonlight, while below were the twinkling lights of the reclamation dam that the government was building. Then I saw the overhead bridge which spanned the river just above the dam. Like a person in a dream I watched the swaying pile-driver, hoping against impending odds that the crane would swing true at the last moment.

But when we were fifty yards from the bridge the great driver crane swung across the track. I yelled at Anderson to jump and I prepared to do the same. But before I could get to the side of the car the crash came and something struck me on the head, and then came darkness. I could dimly sense that I was falling and I supposed that the end of life had come.

The next thing that I knew was when I was revived by plunging into the cold waters of the reservoir. I then started the fight for life and at last got out from under the wreckage, but I was blowing like a porpoise. I then took note of my injuries and was relieved to find that I had no broken bones, and, with the exception of a few minor cuts and bruises, was uninjured.

I then heard Anderson calling for help. I reached him about the same time that the men from the reclamation forces did, and we found him in the water, holding up his wife, who had fainted, and who was nearly drowned. His greatest anxiety was for the safety of his baby girl and he begged us to look for her.

Willing hands searched the wreckage and at last got inside the wrecked cook-car, where we found the baby girl in one of the wrecked bunks, still rolled up in the mattress.

She was uninjured, but was crying fitfully and trying to get out. We were none too soon, as the cook-car settled into the water a few moments after we had taken little Hazel out.

Anderson and his wife and baby were taken to the reclamation camp and there made comfortable. As soon as I reached the camp I called Meadows station and asked the agent about the fate of the rest of my train. I was greatly relieved to learn that they had got the train under control about five miles down the canyon. They were glad to learn that we were all safe. It was a good thing that the bridge men were in the cars ahead of the driver or some of them would most likely have been killed or crippled.

Thankful that his family had been spared, Anderson never took them with him again, but hired a Chinaman to do the cooking.

W. H. Henry, "The Wreck of the Pile-Driver," *The Railway Conductor* 32, no. 4 (April 1915): 252–54.

THE HOBOES' PENDULUM OF DEATH

Two bindle stiffs.

By the turn of the century thousands of men and boys were "riding the rails" in search of adventure, employment, or a way of life. They were hoboes, a term whose origin has never been satisfactorily explained. Every tank town and big city in America had to deal with the problem of these rail-borne drifters. From the standpoint of the general public, they were thieves at best. At worst they were violent criminals.

Of all the bindle stiffs, yeggs, and tramps that ever rode the rails, none was so famous and successful in his craft than Leon Ray Livingston (1872–1944), better known as "A-No.1—the Rambler." Livingston was such an accomplished hobo that he wrote twelve books about the life, always advising young readers not to follow in his path. In one book he stated that his object "was to prove to boys and men of restless dispositions, that by their heeding of the 'Lure of the Wanderlust,' they not only wreck their own futures, but very often the lives and happiness of their parents, and sometimes those of their families as well."

Leon Ray Livingston, from *Trail of the Tramp*, 1913.

Still, Livingston never equivocated when it came to his own dangerous style of living. He traveled, he said, because it had become a habit, and he could not stop it. Livingston kept a notebook of autographed comments from notables of his day, including three presidents, and he was sought after for interviews by several newspapers and magazines.

In one of his books, Livingston tells of an occasion in New York, in 1894, when he placed an ad challenging anyone to accompany him on a coast-to-coast tramp. After receiving an "avalanche of responses," Livingston records that he chose an 18-year-old former sailor who wanted to visit his family in California before leaving on an around-the-world adventure.

Taken by the youth's easy, straightforward ways, Livingston agreed to the partnership. Thus they set off on their journey to the West Coast. The young accomplice earlier had earned the tramp moniker of "Cigaret," but Livingston preferred to call him by his real name, Jack London.

If a portion of following account seems vaguely familiar, it is because the episode was incorporated the 1973 film, Emperor of the North, *starring Lee Marvin, which was loosely based on Livingston's hobo life—*

While we walked on the grade which steeply rose from the banks of the Cuyahoga Creek, the pride of the Clevelanders, a passenger train overtook us. Near the summit of the grade we boarded a passing freight train. While the train stopped at Port Clinton, we went to a residence located nearby to ask for a drink of water wherewith to quench our thirst.

Repeated ringing of the door bell at the front entrance brought no response. Then we tackled the side door with no better result. And though we knocked for some time at the kitchen entrance none came to attend to our want. Deciding that no one was at home, we helped ourselves to our needs at a pump we espied in the back yard. Then we retraced our way to the tracks. There, while we patiently waited for the train to resume its journey, we were nabbed by a constable.

"I want you fellows on a charge of being dangerous and suspicious characters!" snarled the John Law when we vehemently protested against the outrage.

But he took no stock whatever in our objections; quite to the contrary, he came back by snapping handcuffs to our wrists. Then he conveyed us to the residence where we had drunk our fill of water. A typical old maid met us at the entrance of the house.

"For sure! They are the lads who tried to burglarize my home, Mr. Officer!" cackled the ancient dame, identifying us. "They attempted to enter here by way of the doors. Failing to gain an entrance, they were wrenching off the handle of yonder yard pump, when they were chased away by the barking of Atkinson's dog."

Explanations were in order. We had almost exhausted our vocabulary for words wherewith to plead our innocence of intentional wrong-doing, when the constable, though most reluctantly, permitted our release from custody.

At this juncture, the freight train began to depart from Port Clinton. An empty boxcar with its doors standing ajar most invitingly beckoned for a continuance of our journey. Posthaste we ran to connect with the open car. But the minion of law and order took after us. Most likely, he saw a chance to work up against us a "case" which could be made to "stick" in court.

Fee-hungry as he was, he ran so close at our heels, that we escaped from his clutches only by a headlong dive from the car by the door which stood open opposite to the one by which we had entered and through which the cop had climbed aboard to capture us. We hurriedly mounted a side ladder of a passing freight car. But to the roofs of the cars went the John Law chasing us and so compelled our return to solid ground. There he raced after us alongside the moving train.

We were pressed so closely by him, that as a last recourse, we swung onto the gunnels beneath a freight car. Fearing the risk of injury, the cop refused to dive under the running car.

He contented himself to trot by the side of our traveling haven of refuge, all the while bawling commands demanding our voluntary surrender.

"Never count your fees until you've got them earned!" derisively sang out Jack London, at the moment when the constable abandoned the foot race with the train which was running at an ever faster rate of speed.

Onward we traveled lazily stretched across the gunnels and enjoying a deserved respite from the strenuous manhunt we had sustained. Quite ignorant of the fact that the members of the train crew had witnessed the fray, we entertained each other with joshing at the expense of the officer whose authority we had put to naught. But the crew—the rulers of the train—were law-fearing folk who doubtlessly looked askance at our wanton defiance of mandates by which they, the railroaders, abided.

The first thing we were to be aware of, we who were riding in the cellar of Hades beneath the jolting car, was to behold how a member of the train crew left the caboose which swung at the tail end of the train and came running forward over the roofs of the cars.

Although we could not see the man who was abroad beyond the constricted arc of our range of vision, we had a means which allowed a close tab on his doings. This merely was a matter of keeping a watch on his shadow to be correctly informed of his designs and whereabouts.

The silhouette of any trainman abroad on the cars while under way was cast groundward by the sun or, if after nightfall, by the moon, or should the night be a moonless or overcast one, then by the rays emitted by the lighted lantern which after dusk was carried by every railroad man employed on trains or trackage. And this day the sun shone from a cloudless sky.

The shadow of the railroader informed us that he was coming forward and that he had abruptly stopped on arriving atop of the boxcar beneath which we had taken lawless passage. He was a brakeman, as this fact was borne out by the hickory brake club he carried.

He descended on a side ladder of our freight car. Arriving at the lowest rung of the ladder, he took a survey of the lower works of the car and only when he had assured himself that he had correctly judged the distance from the caboose to our hiding place, he yelled: "The conductor of this train has ordered that you get out from under this train. Right now! Instantly! Do you hoboes understand!"

"Get us out from under this speeding train, if you can, sir!" the brakeman was dared by Jack London who was cocky from having defeated the designs of the Port Clinton police officer.

Jack London and Leon Ray Livingston on the front cover of Livingston's 1917 book.

The railroader neither heeded London's tart invitation nor uttered a syllable in reply. But almost instantly the color of his countenance turned to a livid crimson—a telltale sign of fury otherwise controlled.

Presently a diabolic grin made an appearance in his face. To us who believed ourselves safely ensconced beneath the car, this hard grin only helped to confirm our belief that no common agency could dislodge us from under the train at least not while the cars continued to race at better than forty miles an hour.

Having received our defy, the brakeman climbed back to the roof of the car. We heartily laughed when we saw by his shadow that he was returning to the caboose. There he remained but a brief while, for presently we noted his coming again forward over the cars. But this time he carried a coil of light rope, judging the gauge by the diameter of its shadow.

On his approaching to where we were, we discerned a coupling link dangling from one end of the rope. The link, weighty and made of wrought iron, was of the pattern used in the days prior to the universal introduction of automatic car coupling devices.

As the railroader had done on his preceding trip, so at this instance, he halted when he had arrived on the roof of our car. We broke into boisterous laughter at the remarks of derision which we passed regarding the helplessness of the shack in the face of our determination to hobo his train in spite of his orders to the contrary. But the very next minute our laughter was superseded by groans.

By merest chance, I glanced at Jack London. His countenance had assumed an ashen-gray overcast. His eyes were protruding. Further, I could hear his teeth clattering. Too, I felt myself shuddering. And there were no end of other mental and physical manifestations to prove that we both were suffering in the agonies of mortal fright.

There was ample occasion for our panic. The shadow play had told how the trainman had uncoiled the rope. Then he had deliberately lowered the coupling link into the canyon formed by the front wall of our car and the rear side of the one coupled ahead.

A faint metallic clinking was heard. It emanated from the heavy link coming in touch with the coupling apparatus of the cars. On our part another throe of most dreadful fright then RIP! CRASH! THUMP! SMASH! came thunder-like detonations due to the contact with the stationary track by the coupling link, which sustained the momentum of the racing cars.

These detonations alternated with crunching, crushing, and splintering which resounded from the havoc wrought to the iron and wood work of the car by the heavy link which was propelled by titanic force to and from the track, thus faithfully copying the motion of a gigantic pendulum wrecking destruction to everything coming within the radius of its swing.

As the brakeman gradually paid out the rope which held the iron weight in check and to its work, at a similar ratio our personal danger increased. Nearer and ever nearer approached the hideous weapon to where we lay huddled against the gunnels' cast iron supports which transverse limited our retreat from the path of the tool of vengeance employed in bygone days by irate railroaders.

I lay farthest from the death-dealing railroad iron. That is, if the width of a human body might be reckoned to be a span worthy of measurement. But on this occasion even this negligible distance made a vast difference in our demeanor confronted as we were by death.

Protected by the body of Jack London from the thick shower of debris, I instantly realized our dire peril. I yelled to London to hurl himself from the moving train irrespective of the consequences of such a leap to the stationary right of way. He neither made a least

A gunnel, or rod, located below the door of this Chicago & Alton boxcar, circa 1905. At least four rods, laid parallel, supported the floor of the car and made for an acceptable, if chancy, platform for a free ride just inches above the rails.

move nor offered a reply but dazedly stared at the ricocheting link; in fact, he was rendered inanimate by terror of the horrible fate which threatened us.

The paying out of the rope had allowed the link to come within a few inches of where Jack London lay helplessly paralyzed with fear. It was then that I collared my mate by his coat, bodily dragged his nerveless body into my grasp and then, fortunately clearing the rail and the pounding wheels, I flung him to the right of way.

Again Providence intervened. The train was thundering over the crest of a high embankment and when I let go of London, he rolled down a grassy slope.

The next instant I was ready to repeat his vault for life. But ere I let go of my hold on the handle of the car's sliding door, I glanced back into the inferno produced by the pendulum of death. Most timely had we accomplished our exit! The flying weight was bending the gunnels as if they were chaff. Exactly overhead of where we had lain huddled, hand-sized splinters were easily ripped off the car box by the cavorting railroad link. Then I leaped a leap with life or death at stake.

I performed a neat line of somersaults and did other acrobatic stunts ere, like Jack London had before me, I, too, was deposited at the foot of the grass covered incline. There we both lay sprawling, but uninjured. But so terrific was the horror we had passed through that it was some while before we could shake off the grip of our experience.

"And say, A-No. 1, didn't we make a hair's breadth escape from the finish of all things mundane?" gasped Jack London when finally he had recovered so far as to connectedly express his thoughts.

"The Road provides its devotees with such a grand array of dangerous entertainment, one chasing the other so close at the heel, that it is but a matter of days for the hobo to reach the end of his lifetime," I commented contemplatively.

"That's so!" he blurted out and then a weak smile spread over his wan face, indicating that he, too, comprehended the absolute hopelessness of the existence we were leading.

• • •

At midnight and soon after we had wearily trudged into Air Line Junction, the division freight train terminal located just beyond the boundary of the city of Toledo, the fair weather which had prevailed for so many weeks abruptly changed to a drizzly rain that held on.

Rain-stormy days and, more especially, such nights as this one was, were ideal time for the hoboing of railroads. Then detectives and other implacable foes of the Wandering Willies have retreated from track and train to their lairs—yard or station offices, or, if overtaken en route, cabooses or engine cabs.

The downpour had assumed torrential proportions when a freight train departed from the yard. We scanned the cars while

A portion of the rail yards in Kansas City's West Bottoms, 1895. Experienced hoboes knew to keep clear of yards, where railroad police were sure to arrest vagrants. Note the open end door, like the one Livingston wrote about, in the single boxcar right of center. (Missouri Valley Special Collections, Kansas City Public Library, Kansas City, Missouri.)

they passed us to find a shelter aboard from the miserable weather. Through the gloom of the night we saw a small end door of a boxcar to be standing ajar. Mounting to the bumpers of the car, we took note by the flickering light of a match we had struck that the contents of the car was lumber. Evidently an amateur had attended to the loading of the cargo, for while the boards were stacked upwards until flush with the ceiling, a large space remained vacant at the side of the car from where we surveyed its interior.

Momentarily the train was gaining speed. The boxcar, but partly loaded, looked most inviting for a ride through the rain-riven night. Without further delay we climbed aboard. Right then a series of tribulations commenced. The door through which we had entered would not shut. Not even when we pulled and pushed at it with might and main. Neither would it budge when we had returned to the bumpers and there repeated our efforts from outside the car. Finally, after we had wasted the last of our matches, the blackness of the night thwarted a successful search for the cause of the clogging of the door.

Crawling back into the car, only too soon we were to become aware that it offered but a most indifferent shelter from the unfriendly elements. In a corner and farthest from the spot where the rain driving in through the open end door splashed to the floor, we pitched our berth. The track was a straightaway one for many miles beyond Toledo. Then came a curve which routed the train to run in a direction which brought the downpour pattering against our cheeks. This, naturally, sharply aroused us from our sleep. We scurried for shelter to another corner.

But soon another curve sent the storm into our new retreat. There were other curves and more changes of our berthing. We gave up all further attempts to snatch a rest when the floor of the car had begun to resemble a miniature pond.

The train made a first halt at Ryan where it stopped to take on water. During this interim in the journey, two tramps came to keep us company. The newcomers had searched the whole length of the train to find shelter. At the next stop another pilgrim of the Road joined our crowd. Later on, where the train entered a siding, five other tramps were added to our hobo club. Others came and some more until no less than a score of bedraggled tourists were squeezing each other in a space which now became very narrow quarters.

One of the rovers carried a flash lantern. While he undertook to search for the fault which prevented the sliding of the door, I recognized him to be a fellow badly wanted by the police. He removed a splinter of wood that had become tightly wedged in the runway. Obviously, it was placed there by a hobo who feared to be trapped by the shutting and fastening of the door.

While jockeying to provide a favorable position for his train at the Butler (Indiana) coal chute, the engineer slammed the brake shoes with such a sudden force against the rims of the wheels of the cars which were provided with automatic brakes that the remainder of the train was given a most terrific jolt. This sudden shock completely disrupted the natural adhesion supplied by heavy weight to the lumber stowed in our car. The hefty boards were hurled forward with a momentum so great, that some of the hoboes were mercilessly wedged against the sides of the freight car.

With others of our fellows who had come through the accident without sustaining serious harm, we extricated ourselves from the tangled mass of jammed timbers and crushed humans. Then we beat a quick escape into the open.

Extraordinarily precipitate was our exit from the boxcar. Actually we fairly fell over each other to be first to reach the right of way, so anxious we were to remove ourselves promptly from the vicinity of the mishap. We feared an interference with our travel plans and other inconveniences should the authorities decide to hold us as witnesses or, and this was likely, to punish us for trespass.

While Jack London and I scurried for cover, we heard ringing through the darkness the piteous cries of the unfortunates whom in their agony we others had shamelessly deserted. Still we went on; we did not care to get mixed up with new trouble. Then, by chance, while we looked back, we saw how a gleam of light brightly lit up the interior of the car we had quit in such cowardly haste. This brought us to our senses. Responding to the urgings of our outraged consciences, we decided to return to the car and help with the rescue of the injured, irrespective of the outcome of such a step.

Although we were running on an errand of mercy, impelled by a natural suspicion to which every hobo is heir, we took every precaution to guard against untoward surprises. Stealthily mounting the bumpers, we peeped into the end door of the lumber car. We discerned neither officers nor railroaders in the freight car. Instead we saw by the light of his flash lantern that the yegg, he who was hunted for by the authorities, was busily working over the injured. He was not offering succor to those who with their own bodies had become the living cushions which had saved him from sharing their fate; quite to the contrary, he was rifling their pockets of the pitiful contents one might expect in possession of penniless hoboes.

Slipping back into the night we hurriedly held a council of war. Well aware that all murderous hobo criminals carried concealed weapons, we decided against giving battle to the degenerate. Instead we ran to the railroad depot which was located nearby.

There we acquainted the night operator with the particulars of the crime. He promptly gathered a posse. But ere the avengers closed in, the yegg had escaped from the car from which soon afterward an ambulance hauled away several loads of battered hoboes.

When the train departed from the coal chute, on account of the hard rain, we climbed back into the lumber car. But this time we crawled atop of the cargo where the shifting of the timber had left an ample space.

But before we allowed ourselves to drop off to sleep, we barricaded our berth in such manner that we were secure against accident or other interference. By taking this precaution we merely proved that we had practically mastered the lesson of not putting a further trust in an adhesion to each other of either the boards of the lumber or the vultures of the Road.

Leon Ray Livingston, *From Coast to Coast with Jack London*, by A-No. 1, the Famous Tramp (Erie, Pa.: A-No. 1 Publishing, 1917), pp. 37–45, 50–54.

Holding Her Down

Jack London on the deck.

Jack London, the writer whom Livingston claimed to have befriended on the road, wrote his own book about tramping, and never mentioned A-No. 1. The reason was simple. Though London had met A-No. 1 in his journeys, and the two hit it off quite well, they never tramped together. Livingston simply had invented the New York to California adventure with London for publicity purposes.

London's more factual autobiographical account of a coast-to-coast odyssey, entitled The Road, *was published in 1907. It recounts the devil-may-care days of his youth—certainly a time misspent, but full of adventure. A chapter from London's book serves as a representative sampling of that journey—*

When I arrived at the depot, I found, much to my disgust, a bunch of at least twenty tramps that were waiting to ride out the blind baggages of the *Overland*. Now two or three tramps on the blind baggage are all right. They are inconspicuous. But a score! That meant trouble. No train crew would ever let all of us ride.

I may as well explain here what a blind baggage is. Some mail cars are built without doors in the ends; hence, such a car is "blind." The mail cars that possess end doors, have those doors always locked. Suppose, after the train has started, that a tramp gets on to the platform of one of these blind cars. There is no door, or the door is locked. No conductor or brakeman can get to him to collect fare or throw him off. It is clear that the tramp is safe until the next time the train stops. Then he must get off, run ahead in the darkness, and when the train pulls by, jump on to the blind again. But there are ways and ways, as you shall see.

When the train pulled out, those twenty tramps swarmed upon the three blinds. Some climbed on before the train had run a car-length. They were awkward dubs, and I saw their speedy finish.

A car with blind ends, circa 1900.

Of course, the train crew was "on," and at the first stop the trouble began. I jumped off and ran forward along the track. I noticed that I was accompanied by a number of the tramps. They evidently knew their business. When one is beating an overland, he must always keep well ahead of the train at the stops. I ran ahead, and as I ran, one by one those that accompanied me dropped out. This dropping out was the measure of their skill and nerve in boarding a train.

For this is the way it works. When the train starts, the shack rides out the blind. There is no way for him to get back into the train proper except by jumping off the blind and catching a platform where the car-ends are not "blind." When the train is going as fast as the shack cares to risk, he therefore jumps off the blind, lets several cars go by, and gets on to the train. So it is up to the tramp to run so far ahead that before the blind is opposite him the shack will have already vacated it.

I dropped the last tramp by about fifty feet, and waited. The train started. I saw the lantern of the shack on the first blind. He was riding her out. And I saw the dubs stand forlornly by the track as the blind went by. They made no attempt to get on. They were beaten by their own inefficiency at the very start. After them, in the line-up, came the tramps that knew a little something about the game. They let the first blind, occupied by the shack, go by, and jumped on the second and third blinds. Of course, the shack jumped off the first and on to the second as it went by, and scrambled around there, throwing off the men who had boarded it. But the point is that I was so far ahead that when the first blind came opposite me, the shack had already left it and was tangled up with the tramps on the second blind. A half dozen of the more skilful tramps, who had run far enough ahead, made the first blind, too.

At the next stop, as we ran forward along the track, I counted but fifteen of us. Five had been ditched. The weeding-out process had begun nobly, and it continued station by station. Now we were fourteen, now twelve, now eleven, now nine, now eight. It reminded me of the ten little Indians of the nursery rhyme. I was resolved that I should be the last little Indian of all. And why not? Was I not blessed with strength, agility, and youth? (I was eighteen, and in perfect condition.) And didn't I have my "nerve" with me? And furthermore, was I not a tramp-royal? Were not these other tramps mere dubs and "gay-cats" and amateurs alongside of me? If I weren't the last little Indian, I might as well quit the game and get a job on an alfalfa farm somewhere.

By the time our number had been reduced to four, the whole train crew had become interested. From then on it was a contest of skill and wits, with the odds in favor of the crew. One by one the three other survivors turned up missing, until I alone remained. My, but I was proud of myself! No Croesus was ever prouder of his first million. I was holding her down in spite of two brakemen, a conductor, a fireman, and an engineer.

And here are a few samples of the way I held her down. Out ahead, in the darkness—so far ahead that the shack riding out the blind must perforce get off before it reaches me—I get on. Very well. I am good for another station. When that station is reached, I dart ahead again to repeat the maneuver. The train pulls out. I watch her coming. There is no light of a lantern on the blind. Has the crew abandoned the fight? I do not know. One never knows, and one must be prepared every moment for anything. As the first blind comes opposite me, and I run to leap aboard, I strain my eyes to see if the shack is on the platform. For all I know he may be there, with his lantern doused, and even as I spring upon the steps that lantern may smash down upon my head. I ought to know. I have been hit by lanterns two or three times.

But no, the first blind is empty. The train is gathering speed. I am safe for another station. But am I? I feel the train slacken speed. On the instant I am alert. A maneuver is being executed against me, and I do not know what it is. I try to watch on both sides at once, not forgetting to keep track of the tender in front of me. From any one, or all, of these three directions, I may be assailed.

Ah, there it comes. The shack has ridden out the engine. My first warning is when his feet strike the steps of the right-hand side of the blind. Like a flash I am off the blind to the left and running ahead past the engine. I lose myself in the darkness. The situation is where it has been ever since the train left Ottawa. I am ahead, and the train must come past me if it is to proceed on its journey. I have as good a chance as ever for boarding her.

I watch carefully. I see a lantern come forward to the engine, and I do not see it go back from the engine. It must therefore be still on the engine, and it is a fair assumption that attached to the handle of that lantern is a shack. That shack was lazy, or else he would have put out his lantern instead of trying to shield it as he came forward. The train pulls out. The first blind is empty, and I gain it. As before the train slackens, the shack from the engine boards the blind from one side, and I go off the other side and run forward.

As I wait in the darkness I am conscious of a big thrill of pride. The overland has stopped twice for me—for me, a poor hobo on the bum. I alone have twice stopped the overland with its many passengers and coaches, its government mail, and its two thousand steam horses straining in the engine. And I weigh only one hundred and sixty pounds, and I haven't a five-cent piece in my pocket!

Again I see the lantern come forward to the engine. But this time it comes conspicuously. A bit too conspicuously to suit me, and I wonder what is up. At any rate I have something else to be afraid of than the shack on the engine. The train pulls by. Just in time, before I make my spring, I see the dark form of a shack, without a lantern, on the first blind. I let it go by, and prepare to board the second blind. But the shack on the first blind has jumped off and is at my heels. Also, I have a fleeting glimpse of the lantern of the shack who rode out the engine. He has jumped off, and now both shacks are on the ground on the same side with me. The next moment the second blind comes by and I am aboard it. But I do not linger. I have figured out my countermove. As I dash across the platform I hear the impact of the shack's feet against the steps as he boards. I jump off the other side and run forward with the train. My plan is to run forward and get on the first blind. It is nip and tuck, for the train is gathering speed. Also, the shack is behind me and running after me. I guess I am the better sprinter, for I make the first blind. I stand on the steps and watch my pursuer. He is only about ten feet back and running hard; but now the train has approximated his own speed, and, relative to me, he is standing still. I encourage him, hold out my hand to him; but he explodes in a mighty oath, gives up and makes the train several cars back.

The train is speeding along, and I am still chuckling to myself, when, without warning, a spray of water strikes me. The fireman is playing the hose on me from the engine. I step forward from the car platform to the rear of the tender, where I am sheltered under the overhang. The water flies harmlessly over my head. My fingers itch to climb up on the tender and lam that fireman with a chunk of coal; but I know if I do that, I'll be massacred by him and the engineer, and I refrain.

At the next stop I am off and ahead in the darkness. This time, when the train pulls out, both shacks are on the first blind. I divine their game. They have blocked the repetition of my previous play. I cannot again take the second blind, cross over, and run forward to the first. As soon as the first blind passes and I do not get on, they swing off, one on each side of the train. I board the second blind, and as I do so I know that a moment later, simultaneously, those two shacks will arrive on both sides of me. It is like a trap. Both ways are blocked. Yet there is another way out, and that way is up.

So I do not wait for my pursuers to arrive. I climb upon the upright ironwork of the platform and stand upon the wheel of the handbrake. This has taken up the moment of grace and I hear the shacks strike the steps on either side. I don't stop to look. I raise my arms overhead until my hands rest against the down-curving ends of the roofs of the two cars. One hand, of course, is on the curved roof of one car, the other hand on the curved roof of the other car. By this time both shacks are coming up the steps. I know it, though I am too busy to see them. All this is happening in the space of only several seconds. I make a spring with my legs and "muscle" myself up with my arms. As I draw up my legs, both shacks reach for me and clutch empty air. I know this, for I look down and see them. Also I hear them swear.

I am now in a precarious position, riding the ends of the down-curving roofs of two cars at the same time. With a quick, tense movement, I transfer both legs to the curve of one roof and both hands to the curve of the other roof. Then, gripping the edge of that curving roof, I climb over the curve to the level roof above, where I sit down to catch my breath, holding on the while to a ventilator that projects above the surface. I am on top of the train—on the "decks," as the tramps call it, and this process I have described is by them called "decking her." And let me say right here that only a young and vigorous tramp is able to deck a passenger train, and also, that the young and vigorous tramp must have his nerve with him as well.

The train goes on gathering speed, and I know I am safe until the next stop—but only until the next stop. If I remain on the roof after the train stops, I know those shacks will fusillade me with rocks. A healthy shack can "dewdrop" a pretty heavy chunk of stone on top of a car—say anywhere from five to twenty pounds. On the other hand, the chances are large that at the next stop the shacks will be waiting for me to descend at the place I climbed up. It is up to me to climb down at some other platform.

Registering a fervent hope that there are no tunnels in the next half mile, I rise to my feet and walk down the train half a dozen cars. And let me say that one must leave timidity behind him on such a *passear*. The roofs of passenger coaches are not made for midnight promenades. And if any one thinks they are, let me advise him to try it. Just let him walk along the roof of a jolting, lurching car, with nothing to hold on to but the black and empty air, and when he comes to the down-curving end of the roof, all wet and slippery with dew, let him accelerate his speed so as to step across to the next roof, down-curving and wet and slippery. Believe me, he will learn whether his heart is weak or his head is giddy.

As the train slows down for a stop, half a dozen platforms from where I had decked her I come down. No one is on the platform. When the train comes to a standstill, I slip off to the ground. Ahead, and between me and the engine, are two moving lanterns. The shacks are looking for me on the roofs of the cars. I note that the car beside which I am standing is a "four-wheeler"—by which is meant that it has only four wheels to each truck. (When you go underneath on the rods, be sure to avoid the "six-wheelers"—they lead to disasters.)

I duck under the train and make for the rods, and I can tell you I am mighty glad that the train is standing still. It is the first time I have ever gone underneath on the Canadian Pacific, and the internal arrangements are new to me. I try to crawl over the top of the truck, between the truck and the bottom of the car. But the space is not large enough for me to squeeze through. This is new to me. Down in

The brake beam, seen below this boxcar, hangs just inches above the rail tops. Brake beams were located in front of and behind each wheel truck.

the United States I am accustomed to going underneath on rapidly moving trains, seizing a gunnel and swinging my feet under to the brake-beam, and from there crawling over the top of the truck and down inside the truck to a seat on the cross-rod.

Feeling with my hands in the darkness, I learn that there is room between the brake-beam and the ground. It is a tight squeeze. I have to lie flat and worm my way through. Once inside the truck, I take my seat on the rod and wonder what the shacks are thinking has become of me. The train gets under way. They have given me up at last.

But have they? At the very next stop, I see a lantern thrust under the next truck to mine at the other end of the car. They are searching the rods for me. I must make my getaway pretty lively. I crawl on my stomach under the brake-beam. They see me and run for me, but I crawl on hands and knees across the rail on the opposite side and gain my feet. Then away I go for the head of the train. I run past the engine and hide in the sheltering darkness. It is the same old situation. I am ahead of the train, and the train must go past me.

The train pulls out. There is a lantern on the first blind. I lie low, and see the peering shack go by. But there is also a lantern on the second blind. That shack spots me and calls to the shack who has gone past on the first blind. Both jump off. Never mind, I'll take the third blind and deck her. But heavens, there is a lantern on the third blind, too. It is the conductor. I let it go by. At any rate I have now the full train crew in front of me. I turn and run back in the opposite direction to what the train is going. I look over my shoulder. All three lanterns are on the ground and wobbling along in pursuit. I sprint. Half the train has gone by, and it is going quite fast, when I spring aboard. I know that the two shacks and the conductor will arrive like ravening wolves in about two seconds. I spring upon the wheel of the handbrake, get my hands on the curved ends of the roofs, and muscle myself up to the decks; while my disappointed pursuers, clustering on the platform beneath like dogs that have treed a cat, howl curses up at me and say unsocial things about my ancestors.

But what does that matter? It is five to one, including the engineer and fireman, and the majesty of the law and the might of a great corporation are behind them, and I am beating them out. I am too far down the train, and I run ahead over the roofs of the coaches until I am over the fifth or sixth platform from the engine. I peer down cautiously. A shack is on that platform. That he has caught sight of me, I know from the way he makes a swift sneak inside the car; and I know, also, that he is waiting inside the door, all ready to pounce out on me when I climb down. But I make believe that I don't know, and I remain there to encourage him in his error. I do not see him, yet I know that he opens the door once and peeps up to assure himself that I am still there.

The train slows down for a station. I dangle my legs down in a tentative way. The train stops. My legs are still dangling. I hear the door unlatch softly. He is all ready for me. Suddenly I spring up and run forward over the roof. This is right over his head, where he lurks inside the door. The train is standing still; the night is quiet, and I take care to make plenty of noise on the metal roof with my feet. I don't know, but my assumption is that he is now running forward to catch me as I descend at the next platform. But I don't descend there. Halfway along the roof of the coach, I turn, retrace my way softly and quickly to the platform both the shack and I have just abandoned. The coast is clear. I descend to the ground on the off-side of the train and hide in the darkness. Not a soul has seen me.

I go over to the fence, at the edge of the right of way, and watch. Ah, ha! What's that? I see a lantern on top of the train, moving along from front to rear. They think I haven't come down, and they are searching the roofs for me. And better than that—on the ground on each side of the train, moving abreast with the lantern on top, are two other lanterns. It is a rabbit-drive, and I am the rabbit. When the shack on top flushes me, the ones on each side will nab me. I roll a cigarette and watch the procession go by. Once past me, I am safe to proceed to the front of the train. She pulls out, and I make the front blind without opposition. But before she is fully under way and just as I am lighting my cigarette, I am aware that the fireman has climbed over the coal to the back of the tender and is looking down at me. I am filled with apprehension. From his position he can mash me to a jelly with lumps of coal. Instead of which he addresses me, and I note with relief the admiration in his voice.

"You son-of-a-gun," is what he says.

It is a high compliment, and I thrill as a schoolboy thrills on receiving a reward of merit.

"Say," I call up to him, "don't you play the hose on me any more."

"All right," he answers, and goes back to his work.

I have made friends with the engine, but the shacks are still looking for me. At the next stop, the shacks ride out all three blinds, and as before, I let them go by and deck in the middle of the train. The crew is on its mettle by now, and the train stops. The shacks are going to ditch me or know the reason why. Three times the mighty *Overland* stops for me at that station, and each time I elude the shacks and make the decks. But it is hopeless, for they have finally come to an understanding of the situation. I have taught them that they cannot guard the train from me. They must do something else.

And they do it. When the train stops that last time, they take after me hot-footed. Ah, I see their game. They are trying to run me down. At first they herd me back toward the rear of the train. I know my peril. Once to the rear of the train, it will pull out with me left behind. I double, and twist, and turn, dodge through my pursuers, and gain the front of the train. One shack still hangs on after me. All right, I'll give him the run of his life, for my wind is good. I run straight ahead along the track. It doesn't matter. If he chases me ten miles, he'll nevertheless have to catch the train, and I can board her at any speed that he can.

So I run on, keeping just comfortably ahead of him and straining my eyes in the gloom for cattle-guards and switches that may bring me to grief. Alas! I strain my eyes too far ahead, and trip over something just under my feet, I know not what, some little thing, and go down to earth in a long, stumbling fall. The next moment I am on my feet, but the shack has me by the collar. I do not struggle. I am busy with breathing deeply and with sizing him up. He is narrow-shouldered, and I have at least thirty pounds the better of him in weight. Besides, he is just as tired as I am, and if he tries to slug me, I'll teach him a few things.

But he doesn't try to slug me, and that problem is settled. Instead, he starts to lead me back toward the train, and another possible problem arises. I see the lanterns of the conductor and the other shack. We are approaching them. Not for nothing have I made the acquaintance of the New York police. Not for nothing, in boxcars, by water tanks, and in prison cells, have I listened to bloody tales of manhandling. What if these three men are about to manhandle me? Heaven knows I have given them provocation enough. I think quickly. We are drawing nearer and nearer to the other two trainmen. I line up the stomach and the jaw of my captor, and plan the right and left I'll give him at the first sign of trouble.

Pshaw! I know another trick I'd like to work on him, and I almost regret that I did not do it at the moment I was captured. I could make him sick, what of his clutch on my collar. His fingers, tight-gripping, are buried inside my collar. My coat is tightly buttoned. Did you ever see a tourniquet? Well, this is one. All I have to do is to duck my head under his arm and begin to twist. I must twist rapidly—very rapidly. I know how to do it; twisting in a violent, jerky way, ducking my head under his arm with each revolution. Before he knows it, those detaining fingers of his will be detained. He will be unable to withdraw them. It is a powerful leverage. Twenty seconds after I have started revolving, the blood will be bursting out of his finger-ends, the delicate tendons will be rupturing, and all the muscles and nerves will be mashing and crushing together in a shrieking mass. Try it sometime when somebody has you by the collar. But be quick—quick as lightning. Also, be sure to hug yourself while you are revolving—hug your face with your left arm and your abdomen with your right. You see, the other fellow might try to stop you with a punch from his free arm. It would be a good idea, too, to revolve away from that free arm rather than toward it. A punch going is never so bad as a punch coming.

That shack will never know how near he was to being made very, very sick. All that saves him is that it is not in their plan to manhandle me. When we draw near enough, he calls out that he has me, and they signal the train to come on. The engine passes us, and the three blinds. After that, the conductor and the other shack swing aboard. But still my captor holds on to me. I see the plan. He is going to hold me until the rear of the train goes by. Then he will hop on, and I shall be left behind—ditched.

But the train has pulled out fast, the engineer trying to make up for lost time. Also, it is a long train. It is going very lively, and I know the shack is measuring its speed with apprehension.

"Think you can make it?" I query innocently.

He releases my collar, makes a quick run, and swings aboard. A number of coaches are yet to pass by. He knows it, and remains on the steps, his head poked out and watching me. In that moment my next move comes to me. I'll make the last platform. I know she's going fast and faster, but I'll only get a roll in the dirt if I fail, and the optimism of youth is mine. I do not give myself away. I stand with a dejected droop of shoulder, advertising that I have abandoned hope. But at the same time I am feeling with my feet the good gravel. It is perfect footing. Also I am watching the poked-out head of the shack. I see it withdrawn. He is confident that the train is going too fast for me ever to make it.

And the train is going fast—faster than any train I have ever tackled. As the last coach comes by I sprint in the same direction with it. It is a swift, short sprint. I cannot hope to equal the speed of the train, but I can reduce the difference of our speed to the minimum, and, hence, reduce the shock of impact, when I leap on board. In the fleeting instant of darkness I do not see the iron handrail of the last platform; nor is there time for me to locate it. I reach for where I think it ought to be, and at the same instant my feet leave the ground. It is all in the toss. The next moment I may be rolling in the gravel with broken ribs, or arms, or head. But my fingers grip the handhold, there is a jerk on my arms that slightly pivots my body, and my feet land on the steps with sharp violence.

I sit down, feeling very proud of myself. In all my hoboing it is the best bit of train-jumping I have done. I know that late at night one is always good for several stations on the last platform, but I do not care to trust myself at the rear of the train. At the first stop I run forward on the off-side of the train, pass the Pullmans, and duck under and take a rod under a day-coach. At the next stop I run forward again and take another rod.

I am now comparatively safe. The shacks think I am ditched. But the long day and the strenuous night are beginning to tell on me. Also, it is not so windy nor cold underneath, and I begin to doze. This will never do. Sleep on the rods spells death, so I crawl out at a station and go forward to the second blind. Here I can lie down and sleep; and here I do sleep—how long I do not know—for I am awakened by a lantern thrust into my face. The two shacks are staring at me. I scramble up on the defensive, wondering as to which one is going to make the first "pass" at me. But slugging is far from their minds.

"I thought you was ditched," says the shack who had held me by the collar.

"If you hadn't let go of me when you did, you'd have been ditched along with me," I answer.

"How's that?" he asks.

"I'd have gone into a clinch with you, that's all," is my reply.

They hold a consultation, and their verdict is summed up in: "Well, I guess you can ride, Bo. There's no use trying to keep you off."

And they go away and leave me in peace to the end of their division.

I have given the foregoing as a sample of what "holding her down" means. Of course, I have selected a fortunate night out of my experiences, and said nothing of the nights—and many of them—when I was tripped up by accident and ditched.

In conclusion, I want to tell of what happened when I reached the end of the division. On single track, transcontinental lines, the freight trains wait at the divisions and follow out after the passenger trains. When the division was reached, I left my train, and looked for the freight that would pull out behind it. I found the freight, made up on a sidetrack and waiting. I climbed into a boxcar half full of coal and lay down. In no time I was asleep.

I was awakened by the sliding open of the door. Day was just dawning, cold and gray, and the freight had not yet started. A "con" (conductor) was poking his head inside the door.

"Get out of that, you blankety-blank-blank!" he roared at me.

I got, and outside I watched him go down the line inspecting every car in the train. When he got out of sight I thought to myself that he would never think I'd have the nerve to climb back into the very car out of which he had fired me. So back I climbed and lay down again.

Now that con's mental processes must have been paralleling mine, for he reasoned that it was the very thing I would do. For back he came and fired me out.

Now, surely, I reasoned, he will never dream that I'd do it a third time. Back I went, into the very same car. But I decided to make sure. Only one side-door could be opened. The other side-door was nailed up. Beginning at the top of the coal, I dug a hole alongside of that door and lay down in it. I heard the other door open. The con climbed up and looked in over the top of the coal. He couldn't see me. He called to me to get out. I tried to fool him by remaining quiet. But when he began tossing chunks of coal into the hole on top of me, I gave up and for the third time was fired out. Also, he informed me in warm terms of what would happen to me if he caught me in there again.

I changed my tactics. When a man is paralleling your mental processes, ditch him. Abruptly break off your line of reasoning, and go off on a new line. This I did. I hid between some cars on an adjacent sidetrack, and watched. Sure enough, that con came back again to the car. He opened the door, he climbed up, he called, he threw coal into the hole I had made. He even crawled over the coal and looked into the hole. That satisfied him. Five minutes later the freight was pulling out, and he was not in sight. I ran alongside the car, pulled the door open, and climbed in. He never looked for me again, and I rode that coal car precisely one thousand and twenty-two miles, sleeping most of the time and getting out at divisions (where the freights always stop for an hour or so) to beg my food. And at the end of the thousand and twenty-two miles I lost that car through a happy incident. I got a "set-down," and the tramp doesn't live who won't miss a train for a set-down any time.

Jack London, *The Road* (New York: Macmillan, 1907), pp. 29–52.

One Night's Adventure

Lest we are lead to believe that all hoboes are merely mischievous youth out to play a game of cat and mouse with the trainmen, the following is presented as evidence to the contrary. The fact was that many tramps were dangerous characters, and the brakeman who happened upon them sometimes found his own life in jeopardy. Such was the case in this particular encounter, which certainly became a brakeman's worst nightmare—

Although many years have passed since I made my last trip as a freight brakeman, the terrible experiences of that night are still as fresh in my mind as though it were but last week that it all happened, and even now, after many years, I sometimes awake in the dead of night to find myself bathed in a cold sweat, after having again lived through those terrible scenes in my dreams. But I'm getting ahead of my story.

As a young boy in the village, my greatest delight was to loaf around the railroad yard and watch the men doing their work. The mysterious signal code that sent the engine ahead or back with its long string of cars, that lined up the switches, that coupled or uncoupled the cars, that shunted this car into the clear, or sent another one down to the other end of the track, with its brakeman on top, ready to stop it at its proper place, and then when the necessary switching was done, the sharp "*toot! toot!*" in answer to the conductor's signal—all held a strong fascination for me, and in my fondest dreams I pictured myself in a conductor's uniform, with a badge on my cap, delivering orders to the engineer, instructing the brakeman to take siding at the next town, and performing the numerous other duties of that functionary, and last but not least, strutting up and down the depot platform in the small towns to the admiring glances of the inhabitants.

Long before I became old enough to go braking I had mastered the signal code, and many days of practice in the yard in helping the men do the switching had made me adept at that kind of work, so that when the superintendent finally did give me a position as brakeman, I was the happiest boy on the entire system.

Then began the long, tedious climb toward the top. First on the extra list, then the regular run, and then the fast time freight run that was the pride of the division.

We were due to leave Basin at nine o'clock in the evening and were due in Detroit, one hundred and fifty-seven miles south, at three o'clock in the morning, and as we made only four stops along the way, this was a favorite train for tramps to steal a ride on and a hard time we had of it trying to keep the train clear of these undesirable passengers.

The summer of '94 had been an exceptionally bad summer; thousands of idle men were tramping through the country, many of them desperate and capable of committing any crime, not even hesitating at murder.

As there had been considerable pilfering of cars along the road, the order had gone forth from the main office that the trains must all be kept clear of tramps on penalty of dismissal, so you may be sure that we were exceptionally vigilant in trying to keep our trains free from tramps.

A fifteen-year-old boy switchman learning the trade in 1915.

The track for three miles out of Basin was up a heavy grade, up which the engine laboriously proceeded, pulling the long string of cars behind it at a slow rate of speed and this was where we drove off the tramps who had boarded the train as we were pulling out of town. Many exciting adventures were met with along this piece of track by the trainmen while in the performance of their duties, as the tramps did not take very kindly to being put off almost before their journey had commenced.

On the night in question we were half an hour late in starting out of town. I was on top of the caboose watching to see that the conductor did not miss the rear end of the train, as it was always customary for the conductor to hand the engineer his orders and then, giving him a signal to go ahead, catch the rear as it went by.

After seeing the conductor safely on the train I started to go ahead and drive off tramps as usual. I had covered about one-third of the train when I met a sight which gave me a decided scare. There, lying full length on the top of the car, their heads pillowed upon the running board, their feet dangling over the edge of the car, lay three of the toughest looking specimens of the desperado tramps with whom it had ever been my misfortune to come in contact.

By the dim light of my lantern I could see that the one nearest me was hatless and coatless. There was a bloody handkerchief tied around his head in the shape of a bandage and his face was streaked with clotted blood, giving him a most cruel and murderous appearance. The one in the middle was also hatless, but as he lay on his side with his face turned from me I could not see his features very plainly, but I judged him to be quite a young man, although the marks of dissipation gave him a prematurely old look. The third man lay flat upon his back, the lower part of his face was covered with a growth of beard heavy enough to hide his evil features, while the upper part of his face was hidden by the cap he wore. All three were sound asleep, or at least I thought they were, but how near right I was you will know very soon.

As I stood there looking down at them and turning over in my mind what to do, whether to wake them up and order them off the train or go back to the caboose and get the conductor to help me, the train gave a sudden jerk nearly throwing me off my feet. Before I could regain my balance I heard a loud yell and at the same time saw the man farthest from me leap to his feet and with incredible swiftness hurl himself at me. The onslaught was so sudden that there was no time to retreat and just time to brace myself when he landed on me. I tried to keep on my feet but the acquired momentum together with the might of his body was too much, and we went down together with a crash, he on top. My lamp was hurled far over my head by the impact of his body, thus leaving me in complete darkness.

But although I was down, the fight was not over by any means. By a superhuman effort I rolled him off and, breaking loose from his grasp, leaped to my feet with the intention of taking to my heels for the caboose. But before I could do so, he had me by the one leg in a vise-like grip. Quickly pulling back my other leg I delivered a vicious kick, landing square in his face, and with a terrible scream he disappeared over the edge of the car roof.

It had all happened so quickly, the lurch of the train, the onslaught, the short, sharp struggle, and the disappearance of the tramp over the edge of the car, that I stood there as though rooted to the spot, unable to comprehend it all, unable to move.

In my excitement I had forgotten the other two men and I suppose that they too had been so surprised by the suddenness of it all that they were momentarily paralyzed. But only for a moment, and then with a roar like an angry bull, they were upon me. I saw them coming through the darkness, side by side, on murder bent. There was no time to run now, and knowing that I was no match for them I reached for my gun, determined to sell my life dearly, if at all. Imagine my horror when I found only an empty pocket. I had left the gun in my coat hanging in the caboose locker.

No time now for vain regrets but barely time to step aside as they hurled themselves at me with outstretched arms, meaning no doubt to throw me bodily from the train. This time, however, I was more fortunate and succeeded in evading their grasp and before they could recover for another rush I had delivered a terrific blow on the face of the one nearest to me and I had the soul satisfying knowledge of seeing him sink to the car roof without a murmur. The other man stumbled over the prostrate form of his partner and in falling forward grasped me in his arms and attempted to throw me down. I tried to break away from him, but in vain, so I rained blows upon his face with all my strength and did finally succeed in breaking his grip on me, then turning around, I started on a run for the caboose.

Going over the top of a rapidly moving train in a dark night with a lantern to light the way is no mean feat, and many trainmen have lost their lives that way, either by being jerked off by a sudden lurch of the car, or by falling between the cars. Imagine then, if you can, the difficulty of going over a train in a dark night on a fast running train without any light, and you will perhaps realize the terrible position I was in. I dared not stop for fear the tramps would be after me, and I knew that it was suicidal to keep on ahead. What was I to do?

I had gone but a short distance when I felt my feet give from under me and I was falling down between the cars. Instantly stretching out my arms I luckily caught the end of the running board and for one heartbreaking moment hung there, slipping gradually away, and was expecting every second to lose my grip as the weight of my body began to tell on the muscles. Luckily for me, just as I thought I could hold on no longer, my foot happened to touch the grab iron on the end of the car. Supporting my weight upon the one foot I managed to reach up with my other hand and catch the roof of the car and by an extra effort pull myself upon the top of the car, where I lay for a minute completely exhausted and too frightened to move, my nerve completely gone.

I dared not go on, dared not remain where I was, and expected every second to see my assailants pounce upon me. Three hairbreadth escapes from death in as many minutes, and the knowledge that to

A fight on a freight train, 1910.

remain where I was meant death, and to go over that train equally dangerous, was too much for my overwrought nerves, and I swooned.

The cool night air revived me almost immediately and, with consciousness restored, my nerve came back again and getting up on my feet I started back toward the rear end, feeling my way carefully from one car to the other, when all of a sudden my ear caught the sound of heavy footsteps on the running board, above the roar of the train.

I turned around instantly, and there, not over two steps away, was the tramp coming after me. I just had time to square myself before he got to me, and just as he got within reach I delivered a terrific blow to his face and then we closed in a life and death struggle.

Now, however, conditions were a little different. With the impact of my fist against the tramp's face, a change came over me and where but a moment before I had been a cowardly weakling fighting for life, I now seemed to have lost all sense of fear and in its place were rage and anger.

Angered by the cowardly ruffian, who so persistently sought my life, and with an exultant shout—a shout that must have lain dormant in my ancestors since the time of the hardy Norsemen—I closed in with the tramp in a death battle. Working my arms back and forth like piston rods, I rained blow after blow upon his unprotected face, heard the squash, squash of the blows as they came into contact with his face and body, and felt the warm blood spatter in my face.

The onslaught was so fierce that the tramp gave ground under the stinging rain of blows, then suddenly letting go his hold of me, turned and tried to retreat. But with a cry of rage I was upon him again, fighting like a demon, the insatiable lust of battle driving me on and on to the bitter end regardless of what that end might be.

Again we closed in and it became a fierce struggle for supremacy. The tremendous pressure under which I had been fighting was fast telling on me. My breath was coming in heartbreaking gasps and I thought my lungs would burst under the enormous pressure, while the muscles in my arms and legs felt as though they were being pierced with red hot irons. I could feel the hot, liquor-befouled breath of the tramp striking me in the face, and knew instinctively that he too was suffering agony.

Now, however, it began to dawn upon me that I was no match for my assailant, who was much larger and heavier than myself, and seemed to possess the strength of a grizzly bear. It was only my agility and my years of experience on top of the cars that enabled me to hold my own as long as I had. I realized now that even though we did succeed in staying on top of the car, it would be only a matter of time until I must succumb to his superior strength. Back and forth we struggled in the dark night, with no one to witness the fight, no one to interfere, no one to lend a helping hand.

Suddenly the tramp raised me off my feet and with one supreme effort hurled me over the edge of the roof, but although the strain nearly tore me loose from the tramp, there was still strength enough in my arms and hands to retain their grip on the tramp's clothes and I swung back on the roof and the fierce struggle was renewed.

Now the tramp adopted different tactics; instead of trying to hurl me from the now fast moving train, he began to reach for my throat. I struggled with the desperation of a madman to keep his hand from reaching my throat and as we swayed back and forth I had barely time to realize that we were on the edge of the roof, when my foot trod the empty air and I felt the sickening sensation of falling through space, then the shock of striking and—blessed oblivion.

When I regained consciousness I was in a bed in a hospital with two of my fellow trainmen standing watch over me. I tried to raise up on my elbow to look around and talk to them, but the severe pain which accompanied the movement caused me to fall back with a cry of agony that brought the nurse to my side instantly, with instruction to lay perfectly still, adding the information that a man with as many broken bones in his body as I had could not think of leaving his bed for some weeks at least.

Several days later I learned from the conductor how he, becoming anxious over my long absence and not seeing my light over the top of the train, armed himself, and with his lantern to light the way, proceeded over the train to the engine, and failing to find me, stopped at the next telegraph station to notify the train dispatcher of my disappearance and requested that the yard crew be sent out to look for me—how a mile from town they found the dead body of the first tramp and three miles farther out at the foot of a steep embankment, they found the other tramp and myself, both unconscious, but still breathing—how they gathered us up and placing us on the car which they had brought along for that purpose, hurried us to the hospital, where the tramp died the next day without having regained consciousness.

The police, upon making an investigation, learned that the three men had been drinking in a saloon the greater part of the afternoon and early evening and then had attempted to hold up the saloon and its customers. The saloon proprietor had opened fire upon the three men and two of them had returned the fire before making a retreat, but fortunately the bullets, all except one, flew wide of the mark. That one bullet had cut a bad gash in the head of the tramp that had given me such a hard fight. Later on they learned that both the men were ex-convicts, having served terms in prison for highway robbery and burglary. The third man got away entirely, and to the best of my knowledge, is still at large, a menace to the law-abiding public.

When I was considered well enough to leave the hospital, the chief of police invited me to call at his office for a friendly visit. At the end of the visit and as I was leaving the office, he handed me a check for a neat little sum of money, which he explained was a reward for the capture of the two tramps, and had been paid by a western express company, who identified the men as two of a gang who had robbed one of their safes the summer before.

I never went back to the railroad. With the money saved up, together with the reward from the express company, I purchased a few acres of land on the edge of the village, and now evenings instead of preparing to go out on a run you'll find me in my little house with the wife and children, thankful indeed for the outcome of that night's terrible adventure.

E.E. Zackrison, "One Night's Adventure," *The Railroad Trainman* 27, no 3 (March 1910): 176-80.

Owsley and the 1601

The railroad engineer of fiction had much in common with the cowboy. Both, for example, often were depicted with a peculiar fondness toward their steeds. It is easy to see how someone might bear affection toward a horse—a living thing, and in some ways a pet. But a steam engine—a grimy, hot amalgamation of iron and water—an object, animated, yet without life—to love such a creation begs further explanation and examination.

Was it force of habit, or something deeper, that would drive a man to an emotional attachment with a mere mechanical contrivance? One can only wonder as we return to Frank Packard's Hill Division for this 1919 tale about an engineer and his beloved machine—His name was Owsley—Jake Owsley—and he was a railroad man before ever he came to Big Cloud and the Hill Division—before ever the Hill Division was even advanced to the blueprint stage, before steel had ever spider-webbed the stubborn Rockies, before the Herculean task of bridging a continent was more than a thought in even the most ambitious minds.

Owsley was an engineer, and he came from the East, when they broke ground at Big Cloud for a start toward the western goal through the mighty range, a comparatively young man—thirty, or thereabouts. Then other years went by, and the steel was shaken down into the permanent right of way that is an engineering marvel today, and Owsley still held a throttle on a through run—just kept growing a little older, that was all—but one of the best of them, for all that—steadier than the younger men, wise in experience, and with a love for his engine that was like the love of a man for a woman.

It's a strange thing, perhaps, a love like that; but, strange or not, there was never an engineer worth his salt who hasn't had it—some more than others, of course—as some men's love for a woman is deeper than others.

With Owsley it came pretty near being the whole thing, and it was queer enough to see him when they'd change his engine to give him a newer and more improved type for a running mate. He'd refuse point-blank at first to be separated from the obsolete engine that was either carded for some local jerk-water, mixed-freight run, or for a construction job somewhere.

"Leave her with me," he'd say to Regan, the master mechanic. "Leave me with her. You can give my run to someone else, Regan, d'ye mind? It's little I care for the swell run; me and the old girl sticks. I'll have nothing else."

But the bluff, fat, big-hearted, good-natured, little master mechanic, knew his man—and he knew an engineer when he saw one. Regan would no more have thought of letting Owsley get away from the *Imperial*'s throttle than he would have thought of putting call boys in the cabs to run his engines.

"H'm!" he would say, blinking fast at Owsley. "Feel that way, do you? Well, then, mabbe it's about time you quit altogether. I didn't offer you your choice, did I? You take the Imperial with what I give you to take her with—or take nothing. Think it over!"

And Owsley, perforce, had to "think it over"—and, perforce, he stayed on the limited run.

Came then the day when changes in engine types were not so frequent, and a fair maximum in machine-design efficiency had been obtained—and Owsley came to love, more than he had ever loved any engine before, his big, powerful, 1600-class racer, with its four pairs of massive drivers, that took the curves with the grace of a circling bird, that laughed in glee at anything lower than a three per cent grade, and tackled the "fives" with no more than a grunt of disdain—Owsley and the 1601, right from the start, clipped fifty-five minutes off the running time of the *Imperial Limited* through the Rockies, where before it had been nip and tuck to make the old schedule anywhere near the dot.

For three years it was Owsley and the 1601; for three years east and west through the mountains—and a smile in the roundhouse at him as he nursed and cuddled and groomed his big flyer, in from a run. Not now—they don't smile now about it. It was Owsley and the 1601 for three years—and at the end it was still Owsley and the 1601. The two are coupled together—they never speak of one on the Hill Division without the other—Owsley and the 1601.

Owsley! Forty years a railroader—call boy at ten—twenty years of service, counting the construction period, on the Hill Division! Straight and upright as a young sapling at fifty-odd, with a swing through the gangway that the younger men tried to imitate; hair short-cropped, a little grizzled; gray, steady eyes; a beard whose color, once brown, was nondescript, kind of shading tawny and gray in streaks; a slim, little man, overalled and jumpered, with greasy, peaked cap—and, wifeless, without kith or kin save his engine, the star boarder at Mrs. McCann's short-order house. Liked by everybody, known by everybody on the division down to the last Polack construction hand; quiet, no bluster about him, full of good-humored fun, ready to take his part or do his share in anything going, from a lodge minstrel show to sitting up all night and playing trained nurse to anybody that needed one—that was Owsley.

Elbow Bend, were it not for the insurmountable obstacles that Dame Nature had seen fit to place there—the bed of the Glacier River on one side and a sheer rock base of mountain on the other—would have been a black mark against the record of the engineering corps who built the station. Speaking generally, it's not good railroad practice to put a station on a curve—when it can be helped. Elbow Bend, the whole of it, main line and siding, made a curve—that's how it got its name. And yet, in a way, it wasn't the curve that was to blame; though, too, in a way, it was—Owsley had a patched eye that night from a bit of steel that had got into it in the afternoon, nothing much, but a patch on it to keep the cold and the sweep of the wind out.

It was the eastbound run, and, to make up for the loss of time a slow order over new construction work back a dozen miles or so had cost him, the 1601 was hitting a pretty fast clip as he whistled for Elbow Bend. Owsley checked just a little as he nosed the curve—the *Imperial Limited* made no stop at Elbow Bend—and then, as the 1601 sort of got her footing, so to speak, on the long bend, he opened her out again, and the storm of exhausts from her short, stubby stack went echoing through the mountains like the play of artillery.

The light of the west-end siding switch flashed by like a scintillating gem in the darkness. Brannigan, Owsley's fireman, pulled his door, shooting the cab and the heavens full of leaping, fiery red, and swung to the tender for a shovelful of coal. Owsley, crouched a little forward in his seat, his body braced against the cant of the mogul on the curve, was "feeling" the throttle with careful hand as he peered ahead through the cab glass. Came the station lights; the black bulk of a locomotive, cascading steam from her safety, on the siding; and then the thundering reverberation as the

1601 began to sweep past a long, curving line of boxes, flats, and gondolas, the end of which Owsley could not see—for the curve.

Owsley relaxed a little. That was right—Extra No. 49, west, was to cross him at Elbow Bend—and she was on the siding as she should be. His headlight, streaming out at a tangent to the curve, played its ray kaleidoscopically along the sides of the string of freights, now edging the roof of a boxcar, now opening a hole to the gray rock of the cut when a flat or two intervened—and then, sudden, quick as doom, with a yell from his fireman ringing in his ears, Owsley, his jaws clamping like a steel trap, flung his arm forward, jamming the throttle shut, while with the other hand he grabbed at the "air."

Owsley had seen it, too—as quick as Brannigan—a figure, arms waving frantically, for a fleeting second strangely silhouetted in the dancing headlight's glare on the roof of one of the boxcars. A wild shout from the man, fluttering, indistinguishable, reached them as they roared by—then the grind and scream of brake-shoes as the "air" went on—the answering shudder vibrating through the cab of the big racer—the meeting clash of buffer plates echoing down the length of the train behind—and a queer obstructing blackness dead ahead ere the headlight, tardy in its sweep, could point the way—but Owsley knew now—too late.

Brannigan screamed in his ear.

"She ain't in the clear!" he screamed. "It's a swipe! She ain't in the clear!" he screamed again—and took a flying leap through the off-side gangway.

Owsley never turned his head—only held there, grimfaced, tight-lipped, facing what was to come—facing it with clear head, quick brain, doing what he could to lessen the disaster, as forty years had schooled him to face emergency. Owsley—for forty years with his record, until that moment, as clean and unsmirched as the day he started as a kid calling train crews back in the little division town on the Penn in the far East! Strange it should come to Owsley, the one man of all you'd never think it would!

The headlight caught it now—seemed to gloat upon it in a flood of blazing, insolent light—the rear cars of the freight crawling frantically from the main line to the siding—then the pitiful yellow from the cupola of the caboose, the light from below filtering up through the windows.

It seared into Owsley's brain lightning quick, but vivid in every detail in a horrible, fascinating way. It was a second, the fraction of a second since Brannigan had jumped—it might have been an hour.

The front of the caboose seemed to leap suddenly at the 1601, seemed to rise up in the air and hurl itself at the straining engine as though in impotent fury at an unwarranted attack. There was a terrific crash, the groan and rend of timber, the sickening grind and crunch as the van went to matchwood—the debris hurtling along the running boards, shattering the cab glass in flying splinters—and Owsley dropped where he stood—like a log.

And the pony truck caught the tongue of the open switch, and, with a vicious, nasty lurch, the 1601 wrenched herself loose from her string of coaches, staggered like a lost and drunken soul a few yards along the ties—and turned turtle in the ditch.

It was a bad spill, but it might have been worse, a great deal worse—a boxcar and the caboose to the junk heap, and the 1601 for the shops to repair fractures—and nobody hurt except Owsley.

But they couldn't make head or tail of the cause of it.

Everybody went on the carpet for it—and still it was a mystery. The main line was clear at the west end of the siding, and the switch was right; everybody was agreed on that, and it showed that way on the face of it—and that was as it should have been. The operator at Elbow Bend swore that he had shown his red, and that it was showing when the *Limited* swept by. He said he knew it was going to be a close shave whether the freight, a little late and crowding the *Limited*'s running time, would be clear of the main line without delaying the express, and he had shown his red before ever he had heard her whistle—his red was showing. The engine crew and the train crew of Extra No. 49, west, backed the operator up—the red was showing.

Brannigan, the fireman, didn't count as a witness. The only light he'd seen at all was the west-end switch light, the curve had hidden anything ahead until after he'd pulled his door and turned to the tender for coal, and by then they were past the station. And Owsley, pretty badly smashed up, and in bed down in Mrs. McCann's short-order house, talked kind of queer when he got around to where he could talk at all. They asked him what color light the station semaphore was showing, and Owsley said white—white as the moon. That's what he said—white as the moon.[1] And they weren't quite sure he understood what they were driving at.

For a week that's all they could make out of it, and then, with Regan scratching his head over it one day in confab with Carleton, the superintendent, it came more by chance than anything else.

"Blamed if I know what to make of it!" he growled.

"Ordinary, six men's words would be the end of it, but Owsley's the best man that ever latched a throttle in our cabs, and for twenty years his record's cleaner than a baby's. What he says now don't count, because he ain't right again yet; but what you can't get away from is the fact that Owsley's not the man to have slipped a signal. Either the six of them are doing him cold to save their own skins, or there's something queer about it."

[1] At the time of this story, a white signal light indicated "clear," much the same as a green signal in present times.

Carleton, "Royal" Carleton, in his grave, quiet way, shook his head.

"We've been trying hard enough to get to the bottom of it, Tommy," he said. "I wish to the Lord we could. I don't think the men are lying—they tell a pretty straight story. I've been wondering about that patch Owsley had on his eye, and—"

"What's that got to do with it?" cut in the blunt little master mechanic, who made no bones about his fondness for the engineer. "He isn't blind in the other, is he?"

Carleton stared at the master mechanic for a moment, pulling ruminatively at his brier; then—they were in the super's office at the time—his fist came down with a sudden bang upon the desk.

"I believe you've got it, Tommy!" he exclaimed.

"Believe I've got it!" echoed Regan, and his hand halfway to his mouth with his plug of chewing stopped in mid-air. "Got what? I said he wasn't blind in the other, and neither he is—you know that as well as I do."

"Wait!" said Carleton. "It's very rare, I know, but it seems to me I've heard of it. Wait a minute, Tommy."

He was leaning over from his chair and twirling the little revolving bookcase beside the desk, as he spoke—not a large library was Carleton's, just a few technical books, and his cherished *Britannica*. He pulled out a volume of the encyclopedia, laid it upon his desk, and began to turn the leaves.

"Yes, here it is," he said, after a moment. "Listen"—and he commenced to read rapidly:

"'The most common form of Daltonism'—that's colorblindness you know, Tommy—'depends on the absence of the red sense. Great additions to our knowledge of this subject, if only in confirmation of results already deduced from theory, have been obtained in the last few years by Holmgren, who has experimented on two persons, each of whom was found *to have one color-blind eye,* the other being nearly normal.'"

"Color-blind!" spluttered the master mechanic.

"In one eye," said Carleton, sort of as though he were turning a problem over in his mind. "That would account for it all, Tommy. As far as I know, one doesn't go color-blind—one is born that way—and if this is what's at the bottom of it, Owsley's been color-blind all his life in one eye, and probably didn't know what was the matter. That would account for his passing the tests, and would account for what happened at Elbow Bend. It was the patch that did it—you remember what he said—the light was white as the moon."

"And he's out!" stormed Regan. "Out for keeps—after forty years. Say, d'ye know what this'll mean to Owsley—do you, eh, do you? It'll be hell for him, Carleton—he thinks more of his engine than a woman does of her child."

Carleton closed the volume and replaced it mechanically in the bookcase.

Regan's teeth met in his plug and jerked savagely at the tobacco.

"I wish to blazes you hadn't read that!" he muttered fiercely. "What's to be done now?"

"I'm afraid there's only one thing to be done," Carleton answered gravely. "Sentiment doesn't let us out—there's too many lives at stake every time he takes out an engine. He'll have to try the color test with a patch over the same eye he had it on that night. Perhaps, after all, I'm wrong, and—"

"He's out!" said the master mechanic gruffly. "He's out—I don't need any test to know that now. That's what's the matter, and no other thing on earth. It's rough, damn rough, ain't it—after forty years?"—and Regan, with a short laugh, strode to the window and stood staring out at the choked railroad yards below him.

And Regan was right. Three weeks later, when he got out of bed, Owsley took the color test under the queerest conditions that ever a railroad man took it—with his right eye bandaged—and failed utterly.

But Owsley didn't quite seem to understand—and little Doctor McTurk, the company surgeon, was badly worried, and had been all along. Owsley was a long way from being the same Owsley he was before the accident. Not physically—that way he was shaping up pretty well, but his head seemed to bother him—he seemed to have lost his grip on a whole lot of things. They gave him the test more to settle the point in their own minds, but they knew before they gave it to him that it wasn't much use as far as he was concerned one way or the other.

There was more than a mere matter of color wrong with Owsley now. And maybe that was the kindest thing that could have happened to him, maybe it made it easier for him since the colors barred him anyway from ever pulling a throttle again—not to understand!

They tried to tell him he hadn't passed the color test—Regan tried to tell him in a clumsy, big-hearted way, breaking it as easy as he could—and Owsley laughed as though he were pleased—just laughed, and with a glance at the clock and a jerky pull at his watch for comparison, a way he had of doing, walked out of Riley's, the trainmaster's office, and started across the tracks for the roundhouse.

Owsley's head wasn't working right—it was as though the mechanism was running down—the memory kind of tapering off. But the 1601, his engine—stuck. And it was train time when he walked out of Riley's office that afternoon—the first afternoon he'd been out of bed and Mrs. McCann's motherly hands since the night at Elbow Bend.

That afternoon, as Owsley, out for the first time, walked a little shakily across the turntable and through the big engine doors into the roundhouse, the 1601 was out for the first time herself from the repair shops, and for the first time since the accident was standing on the pit, blowing from a full head of steam, and ready to move out and couple on for the mountain run west, as soon as the Imperial Limited came in off the Prairie Division from the East.

Paxley, big as two of Owsley, promoted from a local passenger run, had been given the *Imperial*—and the 1601. He was standing by the front-end, chatting with Clarihue, the turner, as Owsley came in.

Owsley didn't appear to notice either of the men—didn't answer either of them as they greeted him cheerily. His face, that had grown white from his illness, was tinged a little red with excitement, and his eyes seemed trying to take in every single detail of the big mountain racer all at once. He walked along to the gangway, his shoulders sort of bracing further back all the time, and then with the old-time swing he disappeared into the cab.

He was out again in a minute with a long-spouted oil can, and, just as he always did, started in for an oil around. Paxley and Clarihue looked at each other. And Paxley sort of fumbled aimlessly with the peak of his cap, while Clarihue couldn't seem to get the straps of his overalls adjusted comfortably. Brannigan, Owsley's old fireman, joined them from the other side of the engine.

None of them spoke. Owsley went on oiling—making the round slowly, carefully, head and shoulders hidden completely at times as he leaned in over the rod, poking at the motion-gear. And Regan, who had followed Owsley, coming in, got the thing in a glance—and swore fiercely deep down in his throat.

Owsley finished his round, and, instead of climbing into the cab through the opposite gangway, came back to the front-end and halted before Jim Clarihue.

An engineer oiling around before a run, 1904.

"I see you got that injector valve packed at last," said he approvingly. "She looks cleaner under the guard plates than I've seen her for a long time, too. Give me the 'table, Jim."

Not one of them answered. Regan said afterward that he felt as though there'd been a head-on smash somewhere inside of him. But Owsley didn't seem to expect any answer. He went on down the side of the locomotive, went in through the gangway, and the next instant the steam came purring into the cylinders, just warming her up for a moment, as Owsley always did before he moved out of the roundhouse.

It was Clarihue then who spoke—with a kind of catchy jerk:

"She's stiff from the shops. He ain't strong enough to hold her on the 'table."

Regan looked at Paxley—and tugged at his scraggly little brown mustache.

"You'll have to get him out of there, Bob," he said gruffly, to hide his emotion. "Get him out—gently."

The steam was coming now into the cylinders with a more businesslike rush—and Paxley jumped for the cab. As he climbed in, Brannigan followed, and in a sort of helpless way hung in the gangway behind him. Owsley was standing up, his hand on the throttle, and evidently puzzled a little at the stiffness of the reversing lever, that refused to budge on the segment with what strength he had in one hand to give to it.

Paxley reached over and tried to loosen Owsley's hand on the throttle.

"Let me take her, Jake," he said.

Owsley stared at him for a moment in mingled perplexity and irritation.

"What in blazes would I let you take her for?" he snapped suddenly, and attempted to shoulder Paxley aside. "Get out of here, and mind your own business! Get out!" He snatched his wrist away from Paxley's fingers and gave a jerk at the throttle—and the 1601 began to move.

The 'table wasn't set, and Paxley had no time for hesitation. More roughly than he had any wish to do it, he brushed Owsley's hand from the throttle and latched the throttle shut.

And then, quick as a cat, Owsley was on him.

It wasn't much of a fight—hardly a fight at all—Owsley, from three weeks on his back, was dropping weak. But Owsley snatched up a spanner that was lying on the seat, and smashed Paxley with it between the eyes. Paxley was a big man physically—and a bigger man still where it counts most and doesn't show—with the blood streaming down his face, and half blinded, regardless of the blows that Owsley still tried to rain upon him, he picked the engineer up in his arms like a baby, and with Brannigan, dropping off the gangway and helping, got Owsley to the ground.

Owsley hadn't been fit for excitement or exertion of that kind—for any kind of excitement or exertion. They took him back to his boarding house, and Doctor McTurk screwed his eyes up over him in the funny way he had when things looked critical, and Mrs. McCann nursed him daytimes, and Carleton and Regan and two or three others took turns sitting up with him nights—for a month.

Then Owsley began to mend again, and began to talk of getting back on the Limited run with the 1601—always the 1601. And most times he talked pretty straight, too—as straight as any of the rest of them—only his memory seemed to keep that queer sort of haze over it—up to the time of the accident it seemed all right, but after that things blurred woefully.

Regan, Carleton, and Doctor McTurk went into committee over it in the super's office one afternoon just before Owsley was out of bed again.

"What d'ye say—h'm? What d'ye say, doc?" demanded Regan.

Doctor McTurk, scientific and professional in every inch of his little body, lined his eyebrows up into a ferocious black streak across his forehead, and talked medicine in medical terms into the superintendent and the master mechanic for a good five minutes.

When he had finished, Carleton's brows were puckered, too, his face was a little blank, and he tapped the edge of his desk with the end of his pencil somewhat helplessly.

Regan tugged at both ends of his mustache and sputtered.

"What the blazes!" he growled. "Give it to us in plain railroading! Has he got rights through—or hasn't he? Does he get better—or does he not? H'm?"

"I don't know, I tell you!" retorted Doctor McTurk. "I don't know—and that's flat. I've told you why a minute ago. I don't know whether he'll ever be better in his head than he is now—otherwise he'll come around all right."

"Well, what's to be done?" inquired Carleton.

"He's got to work for a living, I suppose—eh?" Doctor McTurk answered. "And he can't run an engine anymore on account of the colors, no matter what happens. That's the state of affairs, isn't it?"

Carleton didn't answer; Regan only mumbled under his breath.

"Well then," submitted Doctor McTurk, "the best thing for him, temporarily at least, to build him up, is fresh air and plenty of it. Give him a job somewhere out in the open."

Carleton's eyebrows went up. He looked at Regan questioningly.

"He wouldn't take it," said Regan slowly. "There's nothing to anything for Owsley but the 1601."

"Wouldn't take it!' snapped the little doctor. "He's got to take it. And if you care half what you pretend you do for him, you've got to see that he does."

"How about construction work with McCann?" suggested Carleton. "He likes McCann, and he's lived at their place for years now."

"Just the thing!" declared Doctor McTurk heartily. "Couldn't be better."

Carleton looked at Regan again. "You can handle him better than any one else, Tommy. Suppose you see what you can do? And speaking of the 1601, how would it do to tell him what's happened in the last month? Maybe he wouldn't think so much of her as he does now."

"No!" exclaimed Doctor McTurk quickly. "Don't you do it!"

"No," said Regan, shaking his head. "It would make him worse. He'd blame it on Paxley, and we'd have trouble on our hands before you could bat an eyelash."

"Yes; perhaps you're right," agreed Carleton. "Well, then, try him on the construction tack, Tommy."

And so Regan went that afternoon from the super's office over to Mrs. McCann's short-order house, and up to Owsley's room.

"Well, how's Jake today?" he inquired, in his bluff, cheery way, drawing a chair up beside the bed.

"I'm fine, Regan," said Owsley earnestly. "Fine! What day is this?"

"Thursday," Regan told him.

"Yes," said Owsley, "that's right—Thursday. Well, you can put me down to take the old 1601 out Monday night. I'm figuring to get back on the run Monday night, Regan."

Regan ran his hand through his short-cropped hair, twisted a little uneasily in his chair—and coughed to fill in the gap.

"I wouldn't be in a hurry about it, if I were you, Jake," he said. "In fact, that's what I came over to have a little talk with you about. We don't think you're strong enough yet for the cab."

"Who don't?" demanded Owsley antagonistically.

"The doctor and Carleton and myself—we were just speaking about it."

"Why ain't I?" demanded Owsley again.

"Why, good Lord, Jake," said Regan patiently, "you've been sick—dashed near two months. A man can't expect to get out of bed after a layoff like that and start right in again before he gets his strength back. You know that as well as I do."

"Mabbe I do, and mabbe I don't," said Owsley, a little uncertainly. "How'm I going to get strong?"

"Well," replied Regan, "the doc says open-air work to build you up, and we were thinking you might like to put in a month, say, with

Bill McCann up on the Elk River work—helping him boss Polacks, for instance."

Owsley didn't speak for a moment, he seemed to be puzzling something out; then, still in a puzzled way: "And then what about after the month?"

"Why then," said Regan, "then"—he reached for his hip pocket and his plug, pulled out the plug, picked the heart-shaped tin tag off with his thumb nail, decided not to take a bite, and put the blackstrap back in his pocket again. "Why then," said he, "you'll—you ought to be all right again."

Owsley sat up in bed.

"You playing straight with me, Regan?" he asked slowly.

"Sure," said Regan gruffly. "Sure, I am."

Owsley passed his hand two or three times across his eyes.

"I don't quite seem to get the signals right on what's happened," he said. "I guess I've been pretty sick. I kind of had a feeling a minute ago that you were trying to sidetrack me, but if you say you ain't, I believe you. I ain't going to be sidetracked. When I quit for keeps, I quit in the cab with my boots on—no way else. I'll tell you something, Regan. When I go out, I'm going out with my hand on the throttle, same as it's been for more'n twenty years. And me and the old 1601, we're going out together—that's the way I want to go when the time comes—and that's the way I'm going. I've known it for a long time."

"How do you mean you've known it for a long time?"

Regan swallowed a lump in his throat, as he asked the question—Owsley's mind seemed to be wandering a little.

"I dunno," said Owsley, and his hand crept to his head again. "I dunno—I just know." Then abruptly: "I got to get strong for the old 1601, ain't I? That's right. I'll go up there—only you give me your word I get the 1601 back after the month."

Regan's eyes, from the floor, lifted and met Owsley's steadily.

"You bet, Jake!" he said.

"Give me your hand on it," said Owsley happily.

And Regan gripped the engineer's hand.

Regan left the room a moment or two after that, and on his way downstairs he brushed the back of his hand across his eyes.

"What the hell!" he growled to himself. "I had to lie to him, didn't I?"

And so, on the Monday following, Owsley went up to the new Elk River road work, and—

But just a moment, we've overrun our holding orders a bit, and we've got to back for the siding. The 1601 crosses us here.

The 1601 was pretty badly shaken up that night in the spill at Elbow Bend, and when they overhauled her in the shops, while they made her look like new, perhaps they missed something down deep in her vitals in the doing of it; perhaps she was weakened and strained where they didn't know she was; perhaps they didn't *get* clean to the bottom of all her troubles; perhaps they made a bad job of a job that looked all right under the fresh paint and the gold leaf. There's nothing superstitious about that, is there?

It was the night at Elbow Bend that Owsley and the 1601 together first went wrong; and both went into hospital together and came out together to the day—the 1601 for her old run through the mountains, and Owsley with no other idea in life possessing his sick brain than to make the run with her. Owsley had a relapse that day—and that day, twenty miles west of Big Cloud, the 1601 blew her cylinder head off. And from then on, while Owsley lay in bed again at Mrs. McCann's, the 1601, when she wasn't in the shops from an endless series of mishaps, was turning the hair gray on a dispatcher or two, and had got most of Paxley's nerve.

But what's the use of going into all the details—there was enough paper used up in the specification repair sheets! Going slow up a grade and around a curve that was protected with ninety-pound guard-rails, her pony truck jumped the steel where a baby carriage would have held the right of way; she broke this, she broke that, she was always breaking something; and rare was the night that she didn't limp into division dragging the grumbling occupants of the mahogany sleepers after her with her schedule gone to smash. And then, finally, putting a clincher on it all, she ended up, when she was running fifty miles an hour, by shedding a driving wheel, and nearly killing Paxley as the rod ripped through and through, tearing the right-hand side of the cab into mangled wreckage—and that finished her for the *Limited* run.

Do you recall that Owsley, too, was finished for the Limited run?

Superstition? You can figure it any way you like—they've got their own notions on the Hill Division.

When the 1601 came out of the shops again after that, the marks of authority's disapprobation were heavy upon her—the gold leaf of the passenger flyer was gone; the big figures on the tender were only yellow paint.

Regan scowled at her as they ran her into the yards.

"Damn her!" said Regan fervently; and then, as he thought of Owsley, he scowled deeper, and yanked at his mustache.

"Say," said Regan heavily, "it's queer, ain't it? Blamed queer—h'm—when you come to think of it?"

And so, while the 1601, disfranchised, went to hauling extra freights, kind of a misfit doing spare jobs, anything that turned up, no regular run any more, Owsley, kind of a misfit, too, without any very definite duties, because there wasn't anything very definite they dared trust him with, went up on the Elk River work with Bill McCann, the husband of Mrs. McCann, who kept the short-order house.

Owsley told McCann, as he had told Regan, that he was only up there getting strong again for the 1601—and he went around on the construction work whistling and laughing like a schoolboy, and happy as a child getting strong again for the 1601!

McCann couldn't see anything very much the matter with Owsley—except that Owsley was happy. He studied the letter Regan had sent him, and watched the engineer, and scratched at his bullet head, and blinked fast with his gray Irish eyes.

"Faith," said McCann, "it's them that's off their chumps—not Owsley. Hark to him singin' out there like a lark! An', bedad, ut's mesilf'll tell 'em so!"

And he did. He wrote his opinion in concise, forceful, misspelled English on the back of a requisition slip, and sent it to Regan. Regan didn't say much—just choked up a little when he read it. McCann wasn't strong on diagnosis.

It was still early spring when Owsley went to the new loop they were building around the main line to tap a bit of the country south, and the chinook, blowing warm, had melted most of the snow, and the creeks, rivers, and sluices were running full—the busiest time in all the year for the trackmen and section hands. It was a summer's job, the loop—if luck was with them—and the orders were to push the work, the steel was to be down before the snow flew again. That was the way it was put up to McCann when he first moved into construction camp, a short while before Owsley joined him.

"Then give me the stuff," said McCann. "Shoot the material along, an' don't lave me bitin' me finger nails for the want av ut—d'ye moind?"

So the Big Cloud yards, too, had orders—standing orders to rush out all material for the Elk River loop as fast as it came in from the East.

The first chance they got in a lull of pressure, they sent the material west with the only spare engine that happened to be in the roundhouse at the time—the 1601—and never thought of Owsley. Regan might have, would have, if he had known it; but Regan didn't know it—then. Regan wasn't handling the operating.

Perhaps, after all, they needn't have been in a belated hurry that day—McCann and his foreigners had done nothing but hug their shanties and listen to the rain washing the ballast away for two days and a half, until, as it got dark on that particular day, barely a week after Owsley had come to the work, they listened, by way of variation, to the chime whistle of an engine that came ringing down with the wind.

McCann and Owsley shared a little shanty by themselves, and McCann was trying to initiate Owsley into the mysteries of that grand old game so dear to the hearts of Irishmen—the game of forty-five. But at the first sound of the whistle, the cards dropped from Owsley's hands, and he jumped to his feet.

"D'ye hear that! D'ye hear that!" he cried.

"An' fwhat av ut?" inquired McCann. "Ut'll be the material we'd be hung up for."

Owsley leaned across the table, his head turned a little sideways in a curious listening attitude—leaned across the table and gripped McCann's shoulders.

"It's the 1601!" he whispered. He put his finger to his lips to caution silence, and with the other hand patted McCann's shoulder confidentially. "It's the 1601!" he whispered—and jumped for the door—out into the storm.

"For the love av Mike!" gasped McCann, staggering to his feet as the lamp flared up and out with the draft.

"Now, fwhat the divil—from this, an' the misfortunate way he picks up forty-foive, mabbe, mabbe I was wrong, an' mabbe ut's queer after all, he is, an'—" McCann was still muttering to himself as he stumbled to the door.

There was no sign of Owsley—only a string of boxes and flats, backed down, and rattling and bumping to a halt on the temporary track a hundred yards away—then the joggling light of a trainman running through the murk and, evidently, hopping the engine pilot, for the light disappeared suddenly and McCann heard the locomotive moving off again.

McCann couldn't see the main line, or the little station they had erected there since the work began for the purpose of operating the construction trains, but he knew well enough what was going on. Off the main line, in lieu of a turntable and to facilitate matters generally, they had built a Y into the construction camp; and the work train, in from the East, had dropped its caboose on the main line between the arms of the Y, gone ahead, backed the flats and boxes down the west-end arm of the Y into the camp, left them there in front of him, and the engine, shooting off on the main line again, via the east-end arm of the Y, would be heading east, and had only to back up the main line and couple on the caboose for the return trip to Big Cloud—there were no empties to go back, he knew.

It was raining in torrents, pitilessly, and, over the gusts of wind, the thunder went racketing through the mountains like the discharge of heavy guns. McCann swore with sincerity as he gazed from the doorway, didn't like the look of it, and was minded to let Owsley go to the devil; but, instead, after getting into rubber boots, a rubber coat, and lighting a lantern, he put his head down to butt the storm, goat fashion, and started out.

"Me conscience 'ud not be clear av anything happened the man," communed McCann, as he battered and sloshed his way along. "Tis wan hell av a night!"

McCann lost some time. He could have made a shortcut over to the main line and the station; but, instead, thinking Owsley might have run up the track beside the camp toward the front-end of the construction train and the engine, he kept along past the string of cars. There was no Owsley; and the only result he obtained from shouting at the top of his lungs was to have the wind slap his voice back in his teeth.

McCann headed then for the station. He took the west-end arm of the Y, that being the nearer to his destination. Halfway across, he heard the engine backing up on the main line, and, a moment later, saw her headlight and the red tail lights of the caboose as she coupled on.

Of course, it was against the rules—but rules are broken sometimes, aren't they? It was a wicked night, and the station, diminutive and makeshift as it was, looked mighty hospitable and inviting by comparison.

The engine crew, Matt Duggan and Greene, his fireman, thought it sized up better while they were waiting for orders than the cab of the 1601 did, and they didn't see why the train crew, MacGonigle, the conductor, and his two brakemen, should have any the better of it—so they left their engine and crowded into the station, too.

There wasn't much room left for McCann when he came in like an animated shower bath. He heard Merle, the young operator—they'd probably been guying him—snap at MacGonigle: "I ain't got any orders for you yet, but you'd better get into the clear on the Y—the Limited, east, is due in four minutes."

"Say!" panted McCann. "Say—" and that was as far as he got. Matt Duggan, making a wild dash for the door, knocked the rest of his breath out of him.

And after Duggan, in a mad and concerted rush, sweeping McCann along with it, the others burst through the door and out on the platform, as, volleying through the storm, came suddenly the quick, staccato bark of engine exhaust.

For a moment, huddled there, trying to get the rights of it, no one spoke—then it came in a yell from Matt Duggan.

"She's *gone*!" he screamed—and gulped for his breath. "She's gone!"

McCann looked, and blinked, and shook the rain out of his face. Two hundred yards east down the track, and disappearing fast, were the twinkling red tail lights of the caboose.

"By the tokens av all the saints," stammered McCann. "Ut's—ut's—" He grabbed at Matt Duggan. "Fwhat engine is ut?"

It was MacGonigle who answered, as they crowded back inside again for shelter—and answered quick, getting McCann's dropped jaw.

"The 1601. What's wrong with you, McCann?"

"Holy Mither!" stuttered McCann miserably. "That settles ut! Ut's Owsley! 'Twas the whistle, d'ye moind—the whistle!"

Merle, young and hysterical, was up in the air.

"The *Limited*! The *Limited*!" he burst out, white faced. "There ain't three minutes between them! She's coming now!"

MacGonigle, grizzled old veteran, cool in any emergency, whirled on the younger man.

"Then stop her!" he drawled. "Don't make a fool of yourself! Show your red and hold her here until you get Big Cloud on the wire—they're both running the *same* way, aren't they, you blamed idiot! Everything's out of the road far enough east of here on account of the *Limited* to give 'em time at headquarters to take care of things. Let 'em have it at Big Cloud."

And Big Cloud got it. Spence, the dispatcher, on the early night trick, got it—and Carleton and Regan, at their homes, got it in a hurried call from Spence over their private keys, that brought them running to headquarters.

"I've cleared the line," said Spence. "The *Limited* is holding at Elk River till Brook's Cut reports Owsley through—then she's to trail along."

Carleton nodded, and took a chair beside the dispatcher's table. Regan, as ever with him in times of stress, tugged at his mustache, and paced up and down the room. He stepped once in front of Carleton and laughed shortly—and there was more in his words, a whole lot more, than he realized then.

"The Lord knows where he'll stop now with the bit in his teeth, but suppose he'd been heading the other way *into* the *Limited*—h'm? Head-on—instead of just tying up all the blamed traffic between here and the Elk—what? We can thank God for that!"

Carleton didn't answer, except by another nod. He was listening to Spence at the key, asking Brook's Cut why they didn't report Owsley through.

The rain rattled at the window panes, and the sashes shook under the gusts of wind; out in the yards below the switch lights showed blurred and indistinct. Regan paced the room more and more impatiently. Carleton's face began to go hard. Spence hung tensely over the table, his fingers on the key, waiting for the sounder to break, waiting for the Brook's Cut call.

It was only seven miles from Elk River, where the stalled passengers of the *Limited*—will you remember this?—grumbled and complained, pettish in their discontent at the delay, only seven miles from there to Brook's Cut, the first station east—only seven miles, but the minutes passed, and still Brook's Cut answered: "No." And Carleton's face grew harder still, and Regan swore deep down under his breath from a full heart, and Spence grew white and rigid in his chair. And so they waited there, waited with the sense of disaster growing cold upon them—waited—but Brook's Cut never reported Owsley "in" or "out" that night.

Owsley? Who knows what was in the poor, warped brain that night? He had heard her call to him, and they had brought him back the 1601, and she was standing there, alone, deserted—and she had called to him. Who knows what was in his mind, as, together, he and the 1601 went tearing through that black, storm-rent night, when the rivers, and the creeks, and the sluices were running full, and the Elk River, that paralleled the right of way for a mile or two to the crossing, was a raging torrent?

Who knows if he ever heard the thundering crash with which the Elk River bridge went out? Who knows, as he swung the curve that opened the bridge approach, without time for any man, Owsley or another, to have stopped, if the headlight playing on the surge of maddened waters meant anything to him? Who knows?

That was where they found them, beneath the waters, Owsley and the 1601—and Owsley was smiling, his hand tight-gripped upon the throttle that he loved.

"I dunno," says Regan, when he speaks of Owsley, "if the mountains out here have anything to do with making a man think harder. I dunno—sometimes I think they do. You get to figuring that the Grand Master mabbe goes a long way back, years and years, to work things out—if it hadn't been for Owsley the Limited would have gone into the Elk that night with every soul on board.

Owsley? That's the way he wanted to go out, wasn't it?—with the 1601. Maybe the Grand Master thought of him, too.

We present now a short bit of mirth about color blindness. One might well expect that this tale made the rounds of many a trackside lunch table in 1896.

Frank L. Packard, *The Night Operator* (New York: A.L. Burt, 1919), pp. 46-76.

The Eye Inspection

As a footnote to the previous story, some Eastern railroads came to favor a method of signaling without reference to color. Known as position light signals, these trackside sentinels used only white or yellow lights. It was the position of the lights—three of them—that provided the information. If lined up vertically, it meant "clear," if lined horizontally, the meaning was "stop," and if diagonal, "approach with caution." But most railroads still relied on the color of signal lights to convey their meaning. Thus, passing a color blindness test was a necessity for any job on the right-of-way.

We present now a short bit of mirth about color blindness. One might well expect that this tale made the rounds of many a trackside lunch table in 1896—

An order was recently issued on a railroad line for all section men to have their eyes examined for color blindness. One of the men failed to show up, but sent a small package neatly tied up and addressed to Roadmaster Hoban for the eye inspector. The inspector found a glass eye wrapped in tissue paper. A hastily scribbled note enclosed read as follows:

"Eye Inspector—Dear Sir: The day before yisterday at none I got word to cum down and have me ise looked after for colur blindness. I had forty-five ties and ten rales to put down beyond the sand cut and my hands were too short to spare me. The right oye that wus first in my head wus put out with the blow of a pick and me glass oye is a perfect figger of the oye that wus not put out, is sent to you. I could spare the glass oye better than the one in the head and if she is culur blind I'll get one that ain't. Yours truly, ANTHONY DRISCOLL."

"Thought His Glass Eye was Color Blind," *Railroad Trainmen's Journal* 13, no. 145 (March 1896): 193.

A Peg-Legged Romance

John Alexander Hill. From *The Historical Register*, 1919.

We arrive now at a story written by John Alexander Hill (1858–1916), a former railroad engineman who turned to far more successful endeavors as a mechanical engineer and publisher. Indeed, Hill is best known for his business legacy—the McGraw-Hill Publishing Company—which began as a technical journal.

And yet for all of his mechanical expertise (he was a close friend of Thomas Edison), Hill displayed a fine talent for storytelling, as we are about to discover. Though purportedly a work of fiction, this tale from 1900 suggests a real life episode through its understated tone and subtle humor. It is a pleasant story, and satisfying to the end. It is...

Some men are born heroes, some become heroic, and some have heroism thrust upon them; but nothing of the kind ever happened to me.

I helped make a hero once—no, I didn't either; I helped make the golden setting after the rough diamond had shown its value.

Miles Diston pulled freight on our road a few years ago. He was of medium stature, dark complexion, but no beauty. He was a manly-looking fellow, well-educated enough, sober, and a steady-going, reliable engineer; you would never pick him out for a hero. Miles was young yet—not thirty— but, somehow or other, he had escaped matrimony; I guess he had never had time. He stayed on the farm at home until he was of age, and then went firing, so that when I first knew him he had barely got to his goal—the throttle.

A good many men, when they first get there, take great interest in their work for a few months—until experience gives them confidence; then they take it easier, look around, and take some interest in other things. Most of them never hope to get above running, and so sit down more or less contented, get married, buy real estate, gamble, or grow fat, each according to the dictates of his own conscience or the inclinations of his make-up. Miles figured a little on matrimony.

I can't explain it; but when a railroad man is in trouble, he comes to me for advice, just as he would go to the company doctor for kidney complaint. I am a specialist in heart troubles. Miles came to me.

Miles was like the rest of them. They don't come right down and say, "Something's the matter with me; what would you do for it?" No, sir! They hem and haw, and laugh off the symptoms, until you come right out and tell them just how they feel and explain the cause; then they will do anything you say. Miles hemmed and hawed a little, but soon came out and showed his symptoms—he asked me if I had ever noticed the "Frenchman's" girl.

"The Frenchman," be it known, was our boss bridge carpenter. He lived at a small place half-way over my division—I was pulling express—and the freights stopped there, changing engines. I knew Venot, the bridge carpenter, very well; met him in lodge occasionally, and once in a while he rode on the engine with me to inspect bridges. His wife was a Canadian woman, and good-looking for her forty years and ten children. The daughter that was killing Miles Diston, Marie Venot, was the eldest, and had just graduated from some sisters' school. She was a very handsome girl, and you could read the romantic nature of her being through her big, round, gray eyes. She was vivacious, and loved to go; but she was a dutiful daughter, and at once took hold to help her mother in a way that made her all the more adorable in the eyes of practical men like Miles.

Miles made the most of his opportunities.

But, bless you, there were other eyes for good-looking girls besides those in poor Miles Diston's head, and he was far from having the field to himself; this he wanted badly, and came to get advice from me.

I advised strongly against wasting energy to clear the field, and in favor of putting it all into making the best show and in getting ahead of all competitors. Under my advice, Miles disposed of some vacant lots, and bought a neat little house, put it in thorough order, and made the best of his opportunities with Marie.

Marie came to our house regularly, and I had good opportunity to study her. She was a sensible little creature, and, to my mind, just the girl for Miles, as Miles was just the man for her. But she had confided to my wife the fact that she never, never could consent to marry and settle down in the regulation, humdrum way; she wanted to marry a hero, someone she could look up to—a king among men.

My wife told her that kings and heroes were scarce just then, and that a lot of pretty good women managed to be comparatively happy with common railroad men. But Marie wanted a hero, and would hear of nothing less.

It was during one of her visits to my house that Miles took Marie out for a ride and (accidentally, of course) dropped around by his new house, induced her to look at it, and told his story, asking her to make the home complete. It would have caught almost any girl; but when Miles delivered her at our door and drove off, I knew that there would be a "For Rent" card on that house in a few days and that Marie Venot was bound to have a hero or nothing.

Miles took his repulse calmly, but it hurt. He told me that Marie was hunting for a different kind of man from him; said that he thought perhaps if he would enlist, and go out to fight Sitting Bull, and come home in a new, brass-bound uniform, with a poisoned arrow sticking out of his breast, she would fall at his feet and worship him. She told him she liked him better than any of the town boys; his calling was noble enough and hard enough; but she failed to see her ideal hero in a man with blue overclothes on and cinders in his ears. If any of Miles' competitors had rescued a drowning child, or killed a bear with a penknife, at this juncture, I'm afraid Marie would have taken him. But, as I have indicated, it was a dull season for heroes.

About this time our road invested in some mogul passenger engines, and I drew one. I didn't like the boiler sticking back

A 2-6-0 wheel arrangement, known as a Mogul engine. From *American Engineer and Railroad Journal*, 1893.

between me and Dennis Rafferty. I didn't like six wheels connected. I didn't like a knuckle joint in the side rod. I didn't like eighteen-inch cylinders. I was opposed to solid-end rods. And I am afraid I belonged to a class of ignorant, short-sighted, bull-headed engineers who didn't believe that a railroad had any right to buy anything but fifteen by twenty-two eight-wheelers—the smaller they were the more men they would want. I got over that a long time ago; but, at the time I write of, I was cranky about it. The moguls were high and short and jerky, and they tossed a man around like a rat in a cornpopper.

One day, as I was chasing time over our worst division, holding on to the armrest and watching to see if the main frame touched the driving-boxes as she rolled, Dennis Rafferty punched me in the small of the back, and said: "Jahn, for the love 've the Vargin, lave up on her a minit. Oi does be chasing that door for the last twinty minits, and not onest has I hit it fair. She's the divil on th' dodge."

Dennis had a pile of coal just inside and just outside of the door, the forward grates were bare, the steam was down, and I went in seven minutes late, too mad to eat—and that's pretty mad for me. I laid off, and Miles Diston took the high-roller out next trip.

Miles didn't rant and write letters or poetry, or marry someone else to spite himself, or take the first steamer for Baraga, or equatorial Africa, as rejected lovers in stories do. It hurt, and he didn't enjoy it, but he bore up all right, and went about his business, just as hundreds of other sensible men do every day. He gave up entirely, however, rented his house, and said he couldn't fill the bill—there wasn't a hero in his family as far back as he could remember.

Miles had been making time with the Black Maria for about a week, when the big accident happened in our town. The boilers in a cotton mill blew up, and killed a score of girls and injured hundreds more. Miles was at the other end of the division, and they hurried him out to take a carload of doctors down. They were given the right of the road, and Miles tested the speed of that mogul—proving that a pony truck would stay on the track at fifty miles an hour, which a lot of us "cranks" had disputed.

A few miles out there is a coaling-station, and at that time they were building the chutes. One of the iron drop-aprons fell just before Miles with the mogul got to it; it smashed the headlight, dented the stack, ripped up the casing of the sand-box and dome, cut a slit in the jacket the length of the boiler, tore off the cab, struck the end of the first car, and then tore itself loose, and fell to the ground.

The throttle was knocked wide open, and the mogul was flying. Miles was thrown down, his head cut open by a splinter, and his foot pretty badly hurt. He picked himself up instantly, and took a look back as he closed the throttle. Everything was "coming" all right, he remembered the emergency of the case, and opened the throttle again. A hasty inspection showed the engine in condition to run—she only looked crippled. Miles had to stand up. His foot felt numb and weak, so he rested his weight on the other foot. He was afraid he would fall off if he became faint, and he had Dennis take off the bell-cord and tie it around his waist, throwing a loop over the reverse lever, as a measure of safety. The right side of the cab and all the roof were gone, so that Miles was in plain sight. The cut in his scalp bled profusely, and in trying to wipe the blood from his eyes, he merely spread it all over himself, so that he looked as if he had been half murdered.

It was this apparition of wreck, ruin, and concentrated energy that Marie Venot saw flash past her father's door, hastening to the relief of the victims of a worse disaster, forty miles away.

Her father came home for his dinner in a few minutes from his little office in the depot. To his daughter's eager inquiry he said there had been some big accident in town and the "extra" was carrying doctors from up the road. But what was the matter with the engine, he didn't know. It was the 170; so it was old man Alexander, he said—and that's the nearest I ever came to being a hero.

Marie knew who was running the 170 pretty well; so after dinner she went to the telegraph office for information, and there she learned that the special had struck the new coal chute at Coalton and that the engineer was hurt. It was time she ran down to see Mrs. Alexander, she said, and that afternoon's regular delivered her in town.

Like all other railroaders not better employed, I dropped round to the depot at train time to talk with the boys and keep track of things in general. The regular was late, but Miles Diston was coming with a special, and came while we were talking about it. Miles didn't realize how badly he was hurt until he stopped the mogul in front of the general office. So long as the excitement of the run was on, so long as he saw the absolute necessity of doing his whole duty until the desired end was accomplished, so long as he had a reputation to protect, his will power subordinated all else. But when several of

A "modern" concrete coaling station. The coal chute dangles above the tracks on the right. A water standpipe, for filling the tender's tank, is pictured in the left foreground. From *Railway Age Gazette*, 1915.

us engineers ran up to the engine, we found Miles hanging to the reverse lever by his safety cord, in a dead faint. We carried him into the depot, and one of the doctors administered some restorative. Then we got a hack and started him and the doctor for my house; but Miles came to himself, and insisted on going to his boardinghouse and nowhere else.

Mrs. Bailey, Miles' boardinghouse keeper, had been a trained nurse, but had a few years before invested in a rather disappointing matrimonial venture. She was one of the best nurses and one of the "crankiest" women I ever knew. I believe she was actually glad to see Miles come home hurt, just to show how she could pull him through.

The doctor found that Miles had an ankle out of joint; the little toe was badly crushed; there was a bad cut in the leg that had bled profusely; there was a black bruise over the short ribs on the right side, and there was a buttonhole in the scalp that needed about four stitches. The little toe was cut off without ceremony, the ankle replaced and hot bandages applied, and other repairs were made, which took up most of the afternoon.

When the doctor got through, he called Mrs. Bailey and myself out into the parlor, and said that we must not let people crowd in to see the patient; that his wounds were not dangerous, but very painful; that Miles was weak from loss of blood, and that his constitution was not in particularly good condition. The doctor, in fact, thought that Miles would be in great luck if he got out of the scrape without a run of fever. Thereafter Mrs. Bailey referred all visitors to me. I talked with the doctor and the nurse, and we all agreed that it would stop most inquisitive people to simply say that the patient had suffered an amputation.

That evening, when I went home, there were two anxious women to receive me, and the younger of them looked suspiciously as if she had been crying. I told them something of the accident, how it all happened, and about Miles' injuries. Both of them wanted to go right down and help "do something," but I told them of the doctor's order and of his fears.

By this time the reporters came; and I called them into the parlor, and then let them pump me. I detailed the accident in full, but declined to tell anything about Miles or his history. "The fact is," said I, "that you people won't give an engineer his just dues. Now, if Miles Diston had been a fireman and had climbed down a ladder with a child, you would have his picture in the paper and call him a hero and all that sort of thing; but here is a man crushed, bleeding, with broken bones, and a crippled engine, who stands on one foot, lashed to his reverse lever, for eighty miles, and making the fastest time ever made over the road, because he knew that others were suffering for the relief he brought."

"That's nerve," said one of the young men.

"Nerve!" said I. "Nerve! Why, that man knows no more about fear than a lion; and think of the sand of the man! This afternoon he sat up and watched the doctor perform that amputation without a quiver; he wouldn't take chloroform; he wouldn't even lie down."

"Was the amputation above or below the knee?" asked the reporter.

"Below." (I didn't state how far.)

"Which foot?"

"Left."

"He is in no great danger?"

"Yes, the doctor says he will be a very sick man for some time—if he recovers at all. Boys," I added, "there's one thing you might mention—and I think you ought to—and that is that it is such heroes as this that give a road its reputation; people feel as though they are safe behind such men."

If Miles Diston had read the papers the next morning he would have died of flattery; the reporters did themselves proud, and they made a whole column of the "iron will and nerves of steel" shown in that "amputation without ether."

Marie Venot was full of sympathy for Miles; she wanted to see him, but Mrs. Bailey referred her to me, and she finally went home, still inquiring every day about him. I don't think she had much other feeling for him than pity. She was down again a week later, and I talked freely of going to pick out a wooden foot for Miles, who was improving right along.

Meanwhile, the papers far and near copied the articles about the "Hero of the Throttle," and the item about the road's interest in heroes attracted the attention of our general passenger agent—he liked the free advertising and wanted more of it—so he called me in one day, and asked if I knew of a choice run they could give Miles as a reward of merit.

I told him, if he wanted to make a show of gratitude from the road, and get a big free advertisement in the papers, to have Miles appointed superintendent of the Spring Creek branch, where a practical man was needed, and then give it out "cold" that Miles had been rewarded by being made superintendent of the road. This was afterwards done, with a great hurrah (in the papers).

The second Sunday after Miles was hurt, Marie was down, and I thought I'd have a little fun with her, and see how she regarded Miles.

"There's quite a romance connected with Diston's affair," said I at the dinner table, rather carelessly. "There is a young lady visiting here in town—I hear she is very wealthy—who saw Miles when we took him off his engine. She sends flowers every day, calls him her hero, and is just crazy for him to get well so she can see him."

"Who is she, did you say?" asked my wife.

"I forgot her name," said I, "but I am here to tell you that she will get Miles if there is any chance in the world. Her father is an army officer, but she says that Miles Diston is a greater hero than the army ever produced."

"She's a hussy," said Marie.

I don't know whether you would call that a bull or a bear movement on the Diston stock, but it went up—I could see that.

A week later Miles was able to come down to our house for dinner, and my wife asked Marie to come also. I met her at the depot, and after she was safe in the buggy, I told her that Miles was up at the house. She nearly jumped out; but I quieted her, and told her she mustn't notice or say a word about Miles' game leg, as he was extremely sensitive about it.

My wife was in the kitchen, and I went to the barn to put out the horse. Marie went to the sitting-room to avoid the parlor and Miles, but he was there, I guess, and Marie found her hero, for when they came out to dinner he had his arm around her.

They were married a month later, and went to Washington, stopping to see us on the way back.

As I came home that night with my patent dinner pail, and with two rows of wrinkles and a load of responsibility on my brow, Marie shook her fist in my face and called me "an old storyteller."

"Storyteller?" said I. "What story?"

"Oh, what story? That *leg* story, of course, you old cheat."

"What leg story?"

"Old innocence; that amputation below the knee—you know."

"Wa'n't it below the knee?"

"Yes, but it was only the little toe."

"John," said Miles, "she cried when she looked for that wooden foot and only found a slightly flat wheel."

"That's just like 'em," said I. "Here Marie only expected a part of a hero, and we give her a whole man, and she kicks—that's gratitude for you."

"I got my hero all right, though," said Marie. "You told me a big fib just the same, but I could kiss you for it."

"Don't you do that," said I. "But if the Lord should send you many blessings, and any of 'em are boys, you might name one after me."

She said she'd do it—and she did.

John A. Hill, *Stories of the Railroad* (Chicago: Jamieson-Higgins, 1900), pp. 77-95.

A Package for Flat Wheel

Perhaps because a brakeman's job was so inherently dangerous, members of that brotherhood often were regarded as a rowdy bunch—crude, fearless, and fun-loving. Conductors, having gained their position through the ranks of brakemen, were only slightly less so, the main difference being that a conductor's personality was somewhat tempered by age and experience.

In either case, those men who walked the swaying car tops and rode in the jerking cabooses shared a disdain for railroad employees who merely manned the stations. They regarded the agents and operators as ornaments—necessary, perhaps, but altogether timid and inferior creatures to be pitied or bullied or made the brunt of jokes, depending upon the situation.

And if an operator committed the unpardonable sin of delaying a train for anything less than an absolute emergency—thereby increasing the length of the road crew's workday—then that deskbound fellow was likely to be taught a lesson, often in the form of a practical joke. The problem with practical jokes is that they sometimes come back to befuddle the jokesters, as in this episode from 1906—

The new superintendent's first official act was to issue the following bulletin to the division: "To all Conductors and brakemen: It is apparent that train crews are disregarding the rule which requires them to carry brake clubs. This order is very necessary on a mountain division and must be adhered to. It has also come to my notice that profanity is excessively practiced, the discontinuance of which is respectfully requested. —J.E. MARTIN, Supt."

A few days after this manifestation of official authority the author of it stepped into caboose No. 273 and greeted pleasantly the line's best known braking crew, Arizona and the "Snake." A brake club profusely decorated with ribbons, hung under a motto which read: "Speak softly, and carry a big stick!"

And the superintendent could not help but smile at the way in which the insistence of his bulletin was carried out by *sundry* and repeated ejaculations of "dear me," and "goodness gracious," carried to an humorous extreme during his introductory conversation with the crew he had heard so much about.

Upon his departure Arizona voiced the sentiments of the crew in a few choice and pointed remarks to the Snake.

"It shore was some effort for y'all to 'speak softly,'" said he. "It sounded all right when I wasn't lookin', but when I see the grotesque maneuvers of that mouth of you'rn wanderin' around over your face like it was huntin' a place to sit down and cry with all them goody words fallin' out of it, it seemed to me about as unnatural as the song of a mockin' bird a-wellin' up from the throat of a crow."

Snake's retort was earnestly rendered in celebration of his return to the language he knew so well, and his pent-up words of natural expression, dammed by his sturdy will, in the presence of the Superintendent, burst forth in a perfect storm of expletives that won the admiration of Arizona and the conductor on the spot.

Preparations for their departure interrupted the discussion and a little later on they slowly pulled out of town.

The first station from the eastern end of the division was in charge of a kid operator, who had come west for the purpose of eradicating from his system a bad case of rheumatism which had given him a very decided limp. For this affliction Snake had promptly dubbed him "Flat Wheel." His office was a boxcar alongside a dry wash, and his immediate surroundings were cactus and coyotes.

Upon the morning in question Flat Wheel turned the red board on No. 39, and when the crew unloaded, gave them a clearance card and a commission to buy for him a couple suits of underclothes in town.

"Just drop 'em off when you come by," he said; "they won't break, and get any color, just so's it's red."

"That kid," said Jim, the conductor, grumbling at the unnecessary delay, "ain't worth his salt. He ought to be 'canned' and give a good man a show."

Arizona and the Snake, thinking of the kid's arrival, when Jim secured him his job and had given him money to tide him over until his first check came, smiled to themselves at the con's transparent efforts to hide his real feelings under a mask of gruffness.

It was night when the terminal was reached and the tired crew pulled down their berths and went to bed.

Jim arose early and was followed some time later by Arizona and the Snake. After breakfast they went up street, Snake to purchase the flannel underwear for the operator, and Jim to perform a self-imposed duty of his own.

Upon his return to the caboose the Snake carried an oblong pasteboard box, and soon after Jim came aboard with his own box, nearly identical to the other.

Then the conductor laboriously began the diction of a letter, and Arizona, letting his eyes rest a moment upon the picture of a girl above the con's head, smiled significantly at Snake.

Later that evening the train began its back trip over the division, by and by arriving at the town where the girl lived. Jim, overwhelmed in a sea of waybills, weight reports, and messages, secured the

Typical caboose interior.

services of the Snake to deliver the letter, along with the box, across the tracks to the young lady's home.

"Be careful and don't crush 'em," said he, "They's flowers."

Upon the return of the Snake they pulled out of town and at their arrival at the next station, signed for the following orders:

"Heavy rains between Hillside and Slagg. All trains will run slow and carefully, and look out for washouts. J.R.L."

This was followed at Hampden with instructions from the trainmaster to examine all bridges before crossing. At the foot of the mountain an engine was ready to push them over the top, and after an hour of creaking, groaning agitation at each end of the train, the summit was reached and the journey down the eastern slope was begun.

They were in the midst of the storm zone, and down the canyons great torrents of water rushed, yellow with foam, snarling at the bridges, washing away anything in their path as they raced for freedom in the valleys far below.

At the little boxcar depot part way down a spot of red in the darkness bade them stop and its frightened operator, Flat Wheel, handed them a message from the dispatcher asking their advice as to the wisdom of "tying up" for the night. A consultation with the engineer was held and a message was sent over both signatures asking for orders and a clear track. Quick as a flash this came back:

"No. 39, Burns, engineer, 1025, has right of track over all trains to Walton. J.R.L."

As Jim handed the order to the engineer and started for the door, a word from Flat Wheel halted him. A dull roaring, easily heard above the storm, fell upon their ears. Flat Wheel sprang to the key and wired the trainmaster at the other end of the line:

"Cloudburst on Hermit's Peak; be here in twenty minutes. Can I leave on No. 39?"

And the answer came back—"Yes."

Five minutes later the train was tearing down grade at a fearful speed, and always at their heels raced the grim Nemesis in the rear. Twenty miles below the track crossed the river, fed by all the drainage on the slope. It would be a race to see what got there first—No. 39 or the raging torrent.

Now and then the reflection of light upon the inky darkness, as the fireman opened and shut the firebox door, revealed to view the figure of Arizona, crouched upon the swaying cars. Every moment the train was gathering momentum, attaining and reaching far beyond the limit of safety.

Flat Wheel was huddled in a corner of the caboose, white-faced and speechless with fear. A sidetrack, with some cars upon it, shot by in a flash, and the impact of sound was like a cannon's roar. Jim sat silent, thinking of the girl, and Snake, eating his lunch in the cupola, pointed a huge piece of pie threateningly at Flat Wheel, and said:

"Pal, if you know how to pray, get busy now. Your friends will soon be sayin' 'how natural you look,' an' there'll be people walkin' slow behind a glass wagon, an' you'll be in it."

Flat Wheel tried to speak but couldn't. He simply cast a reproachful look in the direction of the imperturbable Snake.

Just then Arizona, who had worked his way back from the engine, climbed into the cupola window, and at a wink and a nod from Snake in the direction of Flat Wheel, advanced toward him and held out his hand.

"Good-bye, old man," he said, "it's shore hard to lose out like this, but it's a hundred to one we're down an' out this time."

And the Snake acquiesced in a deep and fervent "Amen."

Flat Wheel's reply was lost in the shudder that shook the jumble of words from his quivering lips, and Jim, at a whisper from Arizona, sank to his knees in an attitude of prayerful solicitation, his head buried in his arms and his shoulders shaking with what looked to be an agony of fear.

Surreptitiously Snake slipped from a rack beside him a fusee, and Arizona reached for the dishpan and the poker.

"We're close to the bridge," shouted Arizona, and springing to the window, he added, in a scream, "My God, we'll never make it!"

Jim sprang to his feet with a yell; the caboose was filled with a jangle of awful noises that crashed into the ears of the frightened operator, and in the midst of the pandemonium a fearful, lurid

A curved wooden trestle, common in turn-of-the-century mountain railroading.

light filled the air, and the banging of the pan and the poker in the hands of Arizona, and the yells of fire from Snake was stilled by the screams of Flat Wheel, which rose above them all.

A moment later they rounded the curve over the bridge into safety on the other side.

A rift of reason seemed to cleave its way into the befuddled brain of their victim and a comprehension of the joke sneaked in. He looked for a moment reproachfully at the grinning trio, then, without a word, pulled down a bunk and went to bed.

A few hours later No. 39 pulled into the terminal. Flat Wheel arose from the bunk and casually began examining the box of underwear that had remained aboard the caboose for the entire trip. Jim had stepped into the yard office to "register in," followed by his brakemen. Moments later they were joined by Flat Wheel, open box in hand, who looked quizzingly at Jim, and said: "I asked for flannels and ye gave me flowers."

A realization of the terrible consequences of the mistake overcame the conductor. Without a word he flung himself out into the night with a parting look of fury at the Snake.

At a little station down the road, the girl, perplexed and enraged, was perusing, for the hundredth time, a letter which read:

"My Very Dear Friend: Please accept this offering as evidence of the sincere regard I have always entertained for you. Though my duties have denied me the pleasure of your companionship, your reception of my advances as I have passed your home gives me hope that you are not adverse to a continuation of my attentions, and so I send these to you to plead my cause.

"I would that mine were the hands to place them on your breast—that mine, the eyes to see you wear them.

"If my suit is acceptable, be at the crossing Wednesday as we go by, and if I see them on you I will be the happiest fellow on the pike. Jim."

The conductor laid off the next trip down and when they passed the crossing Arizona and the Snake energetically waved at a girl that wasn't there anymore.

H.M. Sweezy, "Arizona and the Snake," *Railroad Trainmen's Journal* 23, no. 3 (March 1906): 238–40.

The Phantom Brakeman

In bygone days, a railroad track could be a lonesome, unsettling place in the middle of the night. Those inclined to otherworldly thoughts might even have called it frightening. And when a train came along, the eerie white probe of the headlamp and the occasional moan of the steam whistle only seemed to reinforce the notion that something unseen and menacing lurked nearby. Best not to look too closely, best not to tarry too long, for there was always the chance of encountering…

"Never saw such a dearth of news," growled the city editor, as he sat at his desk beating the arm of his chair with his blue pencil. "We've got to have a story of some description for tomorrow's issue—Jo, can't you get even with those fellows for that 'scoop' of theirs yesterday? You'll have to turn the trick on them to recover what I should term a bit of lost reputation."

I replied in the affirmative to this despotic chief—the czar of the editorial staff—and proceeded to do his bidding. It was raining without, and I donned my heavy ulster preparatory to starting on a still hunt in the dead hours of the night after a piece of news.

The position of reporter on a Rocky Mountain paper in those days was anything but a rosy one, and to start out on such a night as this was calculated to dampen the ardor of the most enthusiastic news gatherer. I will confess that I was somewhat ruffled at the city editor's language to me, and would have sulked had not the added "lost reputation" he mentioned made me angry. This anger served to bring out the resolution of my better self to find something, that very something to be of such a creditable nature that the office force would look upon me with pride in place of the disdain which had been so noticeable in the past twenty-four hours.

I turned down a side street in the "red light" district, and came to a halt at the corner of the only drug store in the neighborhood. I was leaning against the wall of the building, out of the glare of the arc light, when a punch in the side aroused me from my mental ramblings. I turned quickly and perceived the huge hulk of a policeman.

"Hello, Mike," said I, as the burly fellow put out his hand. "You've frightened me out of a week's growth. What do you know? Anything doing around here or down at the yards?" I knew he was a great favorite about the roundhouse and among the switchmen.

"Nothin' goin' on aroun' here," he replied, "but I can steer you up ag'inst an item if you'll take yourself to the roun'house, an' ask for ould man Jack Flanigin, the engine driver. He coom acrost a mishap lasht week, an' it sounds more like a witch tale; but he is sthraight as a yardstick an' you can bank on his nar'tive. You might find him goin' out on his run or coomin' in. He works only one way out of here, an' handles the fast mail goin' west an' the express coomin' back. If you see the engine 77 anywhere aroun' the yards, you can bet on havin' yure man within aisy reach."

"Thank you, Mike," said I, pressing his ponderous hand, "I'll go down there and look into the matter. I know Watson, the night hostler, and if there is anything needed in the way of assistance in locating Flanagan, I am sure he will lend a helping hand."

It was a little past eleven now, and it would take quick work to get the story and have it in cold type ready for the morning issue.

Currier & Ives depiction of a night scene at a railroad yard, circa 1885.

The fast mail was due to leave the Trigg street station at 12:55 a.m. They changed engines at this depot, and if the 77 was to make the trip out tonight on the westward run, I would find the engine and her master at the roundhouse. I quickened my pace at the sight of the first switch lamp and never halted until I reached the yards and came into the network of tracks and "frogs" and countless numbers of switches.

It had turned colder since I left the office and the rain had changed into a fine sleet. There was a thin crust of it over the earth which made walking without overshoes extremely hazardous. This was a night to be feared by the veteran railroader. Many a switchman had fallen between cars during such weather as this, and was picked up by the ambulance corps and taken to the company's hospital.

I found Watson smoking in his little dingy office. Watson's name appeared on the payroll as "night hostler," but in reality he was what you could safely term the night master mechanic. He was a good friend of mine, and I could always rely upon the information he gave me. He would save an item for me every time, and then do his utmost to throw the other force of reporters off the track.

"What's up?" I asked, after I had pulled a stool around back of the stove and was comfortably seated. "Anything in the wind lately?"

"Nothing today," he returned between puffs. "Did you hear about the 'Old Man' and his spook down at Dividing Ridge last Tuesday night?"

"No, I did not; tell me about it."

"Can't do it; wouldn't know how to start about it. You wait here for him; he's due out on the fast mail and will be around in half an hour. The 77, his pranky old steed, is in there now being groomed and waiting for him. Every now and then, she snorts out like she's getting restless," said he, pointing through the grimy window at a massive locomotive. Men were working like beavers on and about the engine, getting her ready for her nocturnal journey.

Watson was telling me about an old sweetheart of his during the days he piloted an engine "east of the river," when the door opened and a swarthy man entered. He wore overalls and a heavy woolen cap with a long visor, from under which peered steel gray eyes that expressed little sentiment and lots of courage. He had the bearing of an engineer of the old school, and such he was.

"Jack," said Watson, "I want you to sit here long enough to tell my friend all about that trouble you had the other night in bringing the express this way. He's one of them news fellows, and I'll warrant that he will put the thing up in such shape it'll flatter you, and make you feel good when you see it in print."

Old man Flanagan grunted his approval and drew forth a piece of tobacco from his jumper pocket. He placed the quid over against a wisdom tooth and settled back in his chair.

"I've had to tell it so many times lately I am getting to be an expert now. I will swear to the truthfulness of it, however, if any one doubts my statement," said he, glancing around in my direction. I saw enough in his expression to satisfy me that he meant more than his words signified.

"To get at the beginning," he continued, "we will have to go back to the time when I was a freight engineer, during the days of the old hand brakes and rowdy railroaders. I had a conductor then who was the meanest, or one of the meanest, men I ever saw. I believe he would have corrupted a preacher just to keep his hand in practice. In his crew was a lad named Grissom, who had a good heart in him, and had received splendid training at home. I knew his aged mother quite well, and she looked to me to keep her only boy in the straight and narrow way.

"Every time we got to the other end of the division, and everybody had washed up and put on his best clothes, Kit would take this boy out and proceed to 'paint the town' as he expressed it. I warned him several times to stay away from the conductor on these drunken debauches, and every time he would promise to swear off; but he went once too often, though.

"One sleety night, as I was going down to get my engine out, I met him in front of the dispatcher's office. He was just recovering from one of his 'painting tours,' and I could see he was still unable to preserve his equilibrium. I told him to ask the train master to send someone out in his place. He said that he could make the trip all right and I left him after telling him to come over and ride on the engine going back. He was braking ahead and I knew this would be a safeguard against his falling from the top of the train.

"We pulled out of the yards with a heavy tonnage just as much as a mogul in the old days could get over those hills. Grissom was sitting on the fireman's box and was half asleep when we left. The heat from the boiler head and the effects of the liquor had put him into a stupor.

"Stranger, have you ever been over that part of the road? Well, Jim here knows what a hard pull it is both ways, and coming east it is considerably steeper. We had a hard time getting to the top that night. My fireman kept the black diamonds in a regular stream from the tender to the furnace door. The drumming of the furnace door and the noise of the exhaust woke Grissom, and he staggered to his feet. He grabbed his lantern and started back over the tender to the train. I caught him by the arm and asked him if he was walking in his sleep.

"'Hell, no,' he growled, 'I am going back on top; you can't hold this heavy train down the Divide with those steam jabs.'

"I saw his determination and released his arm. He climbed back over the coal, and I never saw his light until he started up the end of the car next to the engine. I watched it and its uncertain swing till we turned over the crest of the mountain, and at Dividing Ridge it was lost from view. I thought he had gone to the middle of the train to set some brakes. He had started that way, but the slippery running boards were too much for him.

"The boys in the cupola of the caboose saw his lantern disappear and, as there was no flatcar in the train, they well knew what it meant. They sent old Hannibal, the Negro who did the cooking for the crew, down the train to stop me. I had an idea there was something wrong, and was on the lookout for the signal. We put the train in the siding at Ola, and went back with the engine and caboose. There wasn't anything due from the east and we had thirty minutes on the fast express from the Pacific Slope.

"We found Grissom's remains between the rails and his lighted lantern near the end of the ties. We put him on a stretcher and took him back to his brokenhearted mother. Not a man would touch that lantern; the fact that it remained lighted had its effect on our superstitious crew. It remained there until someone passing fancied it, or it was picked up by the section gang.

"You may look at me in a puzzled way, stranger, and wonder what relation that bears to my part of the story and the accident I encountered; but in the end you will see, and then you will pardon me for my long introductory."

"That is all right," said I, looking admiringly at this eloquent old mechanic. "Listening to stories is my stock in trade. You haven't tired me in the least; in fact, I always enjoy a narrative of the 'Road.'"

"I must hasten with my part of it," said he, as he filled his pipe and lighted it with a wisp of paper. "It is getting late, and I will have to make another trip through sleet and snow, such as I have done for the past thirty winters. I have been through all kinds of weather and have had many different kinds of wrecks, but the incident a few nights ago eclipses the whole kit of my railroad experiences.

"As I was oiling around my engine under the train shed at Denver last Tuesday night, Johnny Gilson, along with the superintendent of the express company, came up to me and inquired 'if the lady could show a clean pair of heels when the occasion demanded.' I asked them to make their meaning clearer.

"'Well,' the superintendent broke in, 'we've got an unusually heavy shipment of gold in the express car, and we have added a couple of guards in case an attempt is made to rob the train. You know,' he concluded, 'it is a ruse of these people to flag a train, pretending trouble ahead, and get you to stop so they can hold you up. What we want you to do is to throw dust in their eyes, when they try these tactics.'

"I can truthfully say that my nerves were unstrung when I mounted my seat box that night. The dangers of a plain everyday run, along with the many responsibilities, are sufficient to try one's

Drifting downhill on a mountain grade, 1892.

steel; but when you add the terrors of being cuffed around with a six-shooter and ordered back to the express car to help dynamite it you are at your row's end.

"The engine worked very well until we got as far out as Pilot Knob. There, she got so she would hardly steam, and no end of laboring with her seemed to do any good. We managed to get to the top of the grade, and then I knew the battle was won. Many miles of down grade were in front of us, with high mountains on the left and deep gorges to the right. Stranger, a magnificent transcontinental train up in that part of God's universe looks like a mere speck. Some of those boulders near Pilot Knob are as big as a tourist sleeper.

"As we started down the mountain in our mad flight, I felt a cold shiver run over me. I turned around to tell my fireman to put on the blower, when I saw a sight that made my blood run cold. There before me, as plain as day, was a ghostlike hand resting on the throttle-bar. I asked my fireman if he saw it, and he shook his head in the negative. No doubt, he was of the opinion that I had lost my reason. I wheeled about to take a look at the rails ahead, and there beyond the few dim lights of Dividing Ridge, I saw a pale white light describe a semicircle above the track. In an instant my hand went over to the air brake to control the train; I knew no one but a railroad man could handle that kind of a signal.

"After I had brought the train to a stop the light disappeared. It all flashed through my mind in an instant—the ghostlike hand and the phantom signal—and I thought of Will Grissom and the night he was killed. I jumped down from my engine and ran forward about two hundred yards to where the light seemed to be. Just about the spot where he was run over I found a rail out. I was looking for the missing piece of steel when the whistling of a bullet made me get under cover. Several of the crew had followed with repeating rifles, and soon had the gang running for the gulches.

"Those scoundrels had pulled that rail out to ditch us. This would have enabled them to get their plunder in a cowardly manner. The only thing that saved us was that signal, and I am firm in the belief that the phantom hand of Grissom did the work. You may not believe in spooks, stranger," said the old man, as he made his way toward the door, "but I do, and when I see one of them around my way again I am going to take off my hat to it."

Jo Custer, "The Phantom Brakeman of Dividing Ridge," *Brotherhood of Locomotive Engineers' Monthly Journal* 40, no. 1 (January 1906): 8-11.

"How Does Yo' Like Yo' Head?"

Smoking car, circa 1900, complete with a barber and spittoons.

There is nothing new under the sun. Some people nowadays are afraid of flying; a century ago the thought of a train ride put many a passenger into near panic. Then, as now, cooler heads sometimes could talk a passenger into facing the fear. But it only took a careless word or two from the wrong person to put the fear right back in place. The thoughtless remark may have been just a harmless phrase, with no intent to cause consternation, such as…

The poor old lady had never ridden in a railroad train before, and now she was making the long journey from New York to Chicago. She asked me for my sympathy.

"My son is in the smoking-place," said she.

"He only laughs at my fears. But I have read of all the horrid accidents in the papers, and I am sure we shall all be plunged into eternity. Are you not afraid? This train goes so fast. I cannot think what keeps it o n the track. My son would take a fast train. 'If you've got to die, you may as well die with a rush,' he says. You would almost think he wanted to be killed. Oh, you only say that to soothe me; but I am not to be deceived. It's reckless to run cars so fast. I know it cannot be done with safety. There! What a lurch! Really, you have taken these trains so often? And did nothing ever happen? And they went so terribly fast, like this? I am sure you ease my mind greatly. I am much obliged to you. I thought it would do me good just to tell how miserable I was. So you have a wife and children, and ain't afraid? I am sure you would not run any risk, and I am glad you comfort me so. There's the colored man. He wants to speak to you."

"Beg yo' pardon, colonel," said the porter. "How does yo' like yo' head?"

"Mercy on me! How do you like your head!' What possesses the man?"

"He means how do I want my berth made up. Make it up with my feet toward the engine, porter, please."

"Oh, I see! Dear me! I'll never dare to go to bed. I shall sit up the whole night, dressed and ready for whatever happens."

"No; don't feel that way. There is no danger. Retire just as you would at home, and you will fall asleep and forget your fears."

"Really? Well, I will follow your advice. You cannot think how you have calmed me."

"I shall undress and sleep like a baby. Porter, leave the window open at the foot of my berth, and leave the screen in."

"Yes, sir. Say, colonel, you's right havin' yo' feet made to'ds de enjyne. Dat's how I allus tell de passengers. 'Feet to'ds de enjyne is de safes' way ebery time,' says I."

"Safest way?" echoed the old lady. "Goodness sakes! How do you mean it's safest?"

"It's easy to see, I kin assure yo', ma'am. If yo's sleepin' feet fust, why dar yo' is; but ef yo're sleepin' wid yo' head to'ds de enjyne, den when dis yer train smashes into some other train, yo' is flung right agin yo' head, an' yo' neck is broke jist like it was a straw."

"Mercy on me! Are we going to smash into some other—"

"No, ma'am ; I didn't say we was a-goin' to. All I say is it's best to be prepared. I've been running on dis yer road twenty-two year, and I've seen 'leven kerlisions, an' every time de folks what's killed is de folks which gits chucked agin their heads. Only last week, in de accident at Osceola, which I were in, a stout lady like you, she— "

But the porter addressed a vacant place. The old lady had fled in search of her son.

Julian Ralph, "How Does Yo' Like Yo' Head?" *Harper's New Monthly Magazine* 87, no. 517 (June 1893): 157.

Chased by an Engine

Taking a cue from the maritime industry, railroads originally gave names to their engines rather than numbers. This quaint practice, which proved impractical from the standpoint of train orders and the sheer quantity of engines, died out long before the end of the nineteenth century.

With that in mind, we present this 1878 story of a predicament. The pickle is that one train inexplicably is chasing another train; and the track ahead isn't clear. Certainly, there will be a wreck! And what of the passengers? Can nothing be done to save their lives? Relax, it's the Gilded Age. There's always a solution, even when you're being…

I was riding on a night train of the Pennsylvania Central from New York to Washington on a mission as newspaper correspondent. We had passed Baltimore, and within an hour's time would be at our place of destination. The conductor had finished collecting the fares, and seeing a vacant seat by my side had dropped into it as if for a little rest at the end of a tiresome day's work. He made an entry in his notebook, closed it, placed it in his breast pocket, buttoned his coat, folded his arms, and then turned to me with a friendly remark, as if now he felt at liberty to lay aside all official dignity and be sociable. I was glad to wile away the time as the train was rushing along in a darkness which concealed all objects of interest without, and so I encouraged the conversation.

"You must have met with some interesting experiences, and perhaps with some great dangers, in the course of your life," said I, the conductor's grizzly beard showing that he might have seen a long service.

"Well, perhaps the most exciting time in my experience was the night I was chased by an engine—a night which this one reminds me of," said he, looking out into the darkness.

"Chased by an engine!" said I, getting interested. "How did that happen?"

"Well," said the conductor, settling down in the cushion and bracing his knees against the back of the seat in front, "many years ago I was running the night express on Long Island from Brooklyn to Greenport, a distance of ninety miles, the entire length of the road. The Long Island road was then a one-horse affair, having only a single track, with switches at the different stations to allow trains to meet and pass. On the evening to which I now refer I started from Brooklyn at ten o'clock with the old *Constitution,* long since broken up, but then the crack engine of the road, with a baggage- or freight-car and three passenger cars. The night was just as dark as a pocket, or, if anything, perhaps a little darker," he added.

"We were the only regular train upon the road that night, with the exception of the Greenport express to Brooklyn, which was to start at ten o'clock and meet us at Lakeland Station, in the middle of the island, switching off there to allow us to pass.

"Well, we were perhaps six or eight miles on our way when I stepped out on the back platform of the rear car to see if it was growing any lighter. We were then going over a part of the road which was as straight as an arrow for a distance of four or five miles. As I was looking back over this stretch I saw behind us, at the distance of three miles or so, what I knew was the headlight of an engine, as it was too bright for anything else; for of course I did not suppose the government had been putting up any lighthouses along the road.

"You may be sure I was a little surprised," said the conductor, "for there wasn't an extra train once a week upon that road, and I knew that there was none going out from Brooklyn that night, anyhow. I waited for a few minutes, until I saw that it was really an engine coming, and, what was more, was gaining rapidly on us, although we were going at our usual rate of speed. When I was satisfied of this fact I hurried forward and said to the engineer, 'Jake, there is a train close behind us.'

"Jake dropped his oil can and his lower jaw at about the same moment, and looked to see whether I was crazy or joking.

"'Well, let the fireman attend to matters here, and come back and see,' said I.

"We hurried to the rear, and in a moment Jake saw as well as myself that if there was any joke in the matter we were the victims of one; and of rather a serious one too, for the train in the rear had gained on us a full mile while I had been forward. The red cinders were pouring out of the smokestack as if from a blast furnace; the headlight threw a glare along the road, burnishing the iron rails to our very wheels.

"Close as he was upon us, the engineer of the advancing train had not given the slightest signal to warn us of his approach, and made no response to our repeated whistles of alarm. He was violating all railroad rules, and if he had determined to secretly run us down he would act just as he was then doing.

"Jake at first seemed to be struck dumb—not so much because he then thought of danger as at the cool impudence of the engineer behind. He looked as if he would like to throttle him. His tongue after a while got in working order, and he broke out: 'What does that crazy fool mean?'

"'The engineer must be either crazy or drunk,' said I. 'If he keeps on in that way ten minutes longer, he will surely be into us.' I signaled the fireman to put on more steam. 'What business the train has upon the road at all tonight is what puzzles me.'

"'I wonder if it isn't an engine the old man is sending down to Jamaica to the shops for repairs?' said Jake. 'I saw the *Ben Franklin* standing on the sidetrack with steam up just as we started. From the way she overhauls us, there can't be much of a train behind her.'

"I did not know but that Jake might be right, for I had seen the *Franklin* standing in the depot when we left. That engine was just as fast as our own, and, if it was without a train attached, as Jake supposed, might easily gain on us, as it seemed to be doing. 'At any rate, we shall see when we pass Jamaica Station whether Jake's theory is correct,' I thought and said to him.

"By this time the fireman, acting as engineer, had given our engine all the steam she would take, and we were slashing along at a lively rate, I tell you," said the conductor. "I was angry enough to have sent a bullet at the crazy engineer following us, and I determined that my first business the next day should be to complain to the superintendent of his foolhardiness. I thought that, possibly, being for the moment his own master and no longer under the immediate orders of a conductor, he was indulging in a kind of a railroad spree, and for a lark was driving us to the top of our speed, expecting to end the race and his day's work at the same time at Jamaica.

"Well, we tore through that sleeping village without stopping long for refreshments, I can assure you, and then Jake and I looked to see our comical friend in the rear pull up at the station and take lodgings for the night. But we were mistaken in our guess. Not a whistle was given by our pursuer as a signal that he intended to stop; not a sign of slackening was shown; but, on the contrary, he was gaining upon us even when we were doing our very best. Sometimes a curve in the road would shut him a moment from our view, but he would round it in an instant, and every new turn brought him more closely upon us.

"Jamaica had been left far behind, and we were out on the wide Hempstead plain. The old *Constitution* was on her muscle. Our train was actually swaying and rocking with speed like a yacht on the waves. The telegraph poles, upon which the light from our windows would glint in the dense darkness, were flying behind us at every second. The sound of our wheels as they struck the ends of the rails was a continuous hum. But, do the best that it might, our engine with its heavy train was no match for the light-weighted one behind that was gaining upon us, and was not the eighth of a mile off.

"The glare from its lantern shone brightly in our faces. I thought Jake's face looked a little pale, and perhaps mine did, too. Now that our pursuer did not halt at Jamaica, we were entirely off our reckonings, and we could make no guess as to the cause of our chase, nor when it would end. The prospect seemed that we might be driven to the end of the road, if we were not overtaken and smashed before it could be reached.

"'That's the *Franklin*, sure,' broke out Jake once more. 'No other engine on the road could overhaul us as we are going now. What can that fool of a Simpson mean by driving her at such a rate? He must be drunk. If the boss don't break him tomorrow he won't get his deserts. He will be into us in two minutes.'

"'You are right, Jake,' said I. 'Go forward and see if you cannot get up a little more headway. Empty a few of those petroleum cans on the wood, and pitch it in and see what can be done.'

"While Jake was forward on his errand I thought over the situation. Here I was with a hundred or two passengers under my care, all ignorant of the danger which I knew they were in. If we should be overtaken and crushed in the rear, the disaster would be a serious one, and would probably cause the death or injury at least of some of the passengers. If we were not smashed in this way, there was another and perhaps a greater danger before us. The train of which I have spoken, which left Greenport when we left Brooklyn, was on its way to meet us on the same track. It should switch off at Lakeland in the middle of the island and allow us to pass an hour after we started, or at eleven o'clock. It was now half-past ten, and we were close to Lakeland already, and would pass there long before the arrival of the Greenport train, which ordinarily got there first. The result would be that we should meet that train beyond Lakeland without warning of our approach, and a collision in front as well as the rear would be the consequence.

"We reached and flew through the Lakeland Depot nearly half an hour ahead of time. Of course the Greenport train was not there yet, but was coming down the road. Our speed was now a little ahead of any ever before made upon the Long Island road. The telegraph poles fairly danced behind us, and the bushes on either side of the track seemed a continuous wall of fire as they were lighted up by the flame which was pouring out of our smokestack. But dangerous as it was for us to keep on, it was just as dangerous to slacken speed, and so on we went."

The conductor rolled his quid from one cheek to the other, raised the window by his side and expectorated into the outer darkness, and became silent for several moments as if burdened by the recollection of his former perils. After waiting a reasonable length of time for him to resume his story, I said, "When the collision occurred, was it with the train in front or in the rear, or with both?"

"Oh, the collision!" said the conductor. "Well, now you come to the ridiculous part of the story. The collision did not take place at all," he said in an apologetic tone, as if there ought to have been a serious accident after so much preparation. " While I was standing on the platform, thinking whether I had better warn the passengers to hold themselves ready for a shock, Jake came from forward dragging after him two large petroleum cans, each of which would hold a quarter of a barrel of oil.

"'Now, then,' said Jake to me, 'if you will oil one side of the track, I will try the other.'

"I saw at once what his plan was. We each brought the mouth of an oil can as near to the polished surface of the rail as possible and commenced pouring the kerosene on it. In less than a minute a half-mile of the iron rails on both sides was nicely oiled and as slippery as the tongue of a Hebrew dealer in second-hand clothes."

"You have raised my expectations of a catastrophe so high that you have been obliged to grease the track so as to let them down again easily," said I, for I felt a little nettled at the unexpected turn the story had taken, and was inclined to believe that the conductor was drawing largely upon his imagination for the facts.

"Why, don't you know that an engine can no more make headway on a greased track than a tomcat can climb a steep roof covered with ice?" said the conductor, with a pitying glance at one so profoundly ignorant of railroad matters as myself. "I slapped Jake on the back, and said, 'Old fellow, your cuteness has brought us all out of a bad scrape.'

"In a few seconds the lantern of the train behind us was getting dim in the distance. We slackened speed and backed down to see 'what the matter was with Simpson,' as Jake said. There stood the old *Ben Franklin* puffing and snorting and pawing like a mad bull, the driving wheels buzzing around on the greased track like all possessed, but not gaining an inch. We sanded the track and bore down upon the old machine.

"Jake was the first aboard, spoiling for a good chance at the engineer, Simpson. But no sign of an engineer, fireman or any other living being was to be found. The engine had only a tender attached, and although there was still a full head of steam on, the fires were getting low.

"We made short work in pushing back to Lakeland. We reached the station, and got fairly upon the switch when the Greenport train, which we should meet there, came in, and we were waiting as if nothing had happened, and as if we had not been fifteen miles out on the road to meet it a few minutes before.

"The telegraph operator at Lakeland handed me a dispatch which read as follows:

"'To CONDUCTOR C—: The *Ben Franklin* has broken loose and is coming up the road. Turn switch at Lakeland and run her off the track. BARTON, *Supt.*'

"To which we replied: 'You see, we did not have much time for turning switches at Lakeland, so we did still better, and saved the old *Ben*—which was not responsible, after all—from a smashup.' E. P. BUFFITT."

E.P. Buffitt, "Chased by an Engine: A Conductor's Story," *Lippincott's Magazine of Popular Literature and Science* 21 (June 1878): 754-57.

A Glossary of Railroad Words and Terms Found in This Anthology

NOTE: The definitions listed below represent the meaning of the word or phrase as it was understood in the context of the stories in this collection. In some cases those meanings have changed with time.

ADJUSTING THE WIRES—Resetting the sensitivity of a telegraph due to variations in line voltage.

AGENT—The principal representative of the railroad at a station. Agents performed many of the same functions as an operator, including telegraphic communications.

ANGLE COCK—A valve used to open or close the air brake line. The angle cock was located at each end of a car, alongside the coupler.

ANNUL AN ORDER—To cancel a previously issued train order.

ANSWER A FLAG—To acknowledge a hand, light, or flag signal to stop, usually with a whistle blast from the engine.

APPLYING THE AIR—Setting the airbrakes throughout the length of the train from a brake lever in the cab.

BALLAST—The roadbed upon which the track and ties sat. Ballast could consist of crushed stone, cinders, or dirt. The ballast drained water away from the track, thus helping to prevent warping or sagging of the track.

BALLOON STACK—A large, flared smokestack on an engine. Balloon stacks were usually found on wood-burning engines. The balloon stack contained baffle plates to inhibit the discharge of large embers, which posed a fire hazard to wooden trestles and other structures, not to mention any bordering forest.

BINDLE STIFF—A hobo, especially one who carried a bedroll or bundle (bindle). The bindle usually was tied on the end of a pole that the hobo slung over his shoulder.

BLOWER—A device on the engine designed to increase the draft of air through the firebox, thereby creating more steam in the boiler.

BLOWING OFF STEAM—When the steam pressure within the boiler exceeded the limit of the safety valve, the valve would open and a heavy rush of steam automatically would escape into the air.

BOILER HEAD—The rear portion of the boiler, which included the opening into the firebox. The engine gauges, valves, and throttle were hung from the boiler head. Also referred to as the "backhead."

BOILER TUBE—See "flue."

BOOMER—A seasonal or temporary railroad employee who traveled from place to place for employment.

BRAKE BEAM—A steel-clad wooden crossbeam that pressed the brake shoes against the wheel surface.

BRAKE CLUB—A short, stout wooden baton used to gain leverage when turning a brake wheel.

BRAKEMAN—A crewmember—usually there were two on a train crew—who coupled and uncoupled the cars, aligned switches, and set the brakes by hand when necessary. Slang terms for a brakeman included brakie, shack, and car-grabber. One brakeman usually rode on the engine, the other in the caboose.

BRAKE WHEEL—A heavy, iron spindle, located at one end of a car, used to manually set the brakes.

BROWNIES—Brownie points, or demerits. The term was derived from a system of rewards and demerits devised by George R. Brown, a railroad executive, in 1886.

BUFFER PLATES—Similar to bumpers (see), buffers were metal plates on the ends of passenger car platforms that absorbed shock caused by slack run-in.

BUMPER—A protruding beam or metal plate on the end of some cars that relieved pressure on the couplers during a sudden braking or hard coupling. The term also was applied to the coupler housing.

BUNCH THE SLACK—All car couplers were inherently loose—enough so as to allow several feet of slack to develop over the length of a train. Bunching the slack—also known as running in the slack or taking up slack—meant forcing the cars closer together. This could be accomplished by braking a moving train and thus taking advantage of momentum to push the rear cars forward, or by backing up the engine against a stopped train.

CABOOSE—Also known as a waycar, van, crummy, or hack, the caboose was the traveling office for the conductor and a vantage point for the brakeman to watch over the train. Besides the conductor's desk and the cupola, the caboose usually contained bunks, an equipment locker, a heating or cooking stove, and a commode. The caboose normally was coupled at the rear of a freight train.

CALL-BOOK—A sign-in register indicating that a trainman was present for duty.

CALLBOY—A railroad employee whose job it was to locate and notify off-duty train crewmen for a coming shift. Sometimes referred to as a "caller."

CAR-GRABBER—A brakeman.

C&E—The term used on train orders to address the conductor and engineer.

CINDER—The fused ash remnant of burnt coal.

CLINKER—An unburnable piece of coal, usually the result of impurities.

CLOSE THE KEY—Disconnect from the telegraph circuit; essentially, to hang up.

COMPOUND—A locomotive with two sets of cylinders on each side.

CONDUCTOR—The crewman in charge of the train. Also called the con, captain, or cap.

CONSOLIDATION—A steam engine that featured two small pilot wheels and eight coupled driving wheels. The consolidation normally was used for freight service.

COPY 3—Make three copies of a train order. One would go to the engineer, one to the conductor, and one for the station files.

COUPLER—The mechanical locking device that held two cars together. The automatic coupler used by the turn of the century, also known as the Janney coupler, resembled the shape and bending motion of the human hand and fingers. See also "link and pin."

COWCATCHER—An attachment located on the front of most engines that was meant to knock any large obstruction from the track. The typical cowcatcher of the period was slanted downward and outward from an angled center ridge.

COUPLING LINK—See "link and pin."

CROWN SHEET—The top, inner portion of the firebox.

CUTTING OFF—Setting the lever (see) to limit the amount of steam entering the cylinder with each piston stroke. Cutting off would save steam and thus increase the boiler pressure.

CYLINDER—The cylindrical-shaped housing located on either side of the engine, in front of the drivers, that contained the piston. It was in the cylinder that steam pressure was converted to motion.

DAY-COACH—A chair car; a passenger car with seats that did not convert to berths for sleeping.

DIAL—The steam pressure gauge.

DIAMONDS—Coal from the tender.

DISPATCHER—A mid-level employee who orchestrated the movement of trains over a division, via telegraphic or telephonic communication with his agents and operators. The dispatcher's job can best be compared to the modern-day air traffic controller.

DISTANT (SIGNAL)—A trackside signal set in advance of the entrance to the territory that it was intended to protect. A distant signal was placed where a curve or other obstruction might have prevented the engineer from seeing the home signal (see) in time to react. The distant signal displayed the same indication as the home signal.

DIVISION—A geographic area under the control of a dispatcher and a superintendent. The length of a division varied according to traffic and density.

DOME—A reference to the steam dome, which was located on top of the boiler. High-pressure steam from the boiler collected in the dome, where the throttle opening was located. The steam dome was further covered with an outer casing.

DOUBLE THE ROAD (OR HILL)—Usually a hill, as series of hills, or a long grade was "doubled" when the engine could not pull the entire train up the grade due to tonnage or steepness. In that case the engine would take approximately half the train up the hill, leave it parked on a parallel track, run back down to the second half of the train, and pull it up the same hill.

DOUBLEHEADER—Two engines assigned to pull a train.

DOWN BRAKES—A term meaning "set the brakes." An engineer could signal for down brakes with one short blast of the whistle.

DOWN IN THE CORNER—Setting the lever (see) in the forward-most notch. This setting allowed steam to enter the cylinders for the full stroke of the piston, which created more power but also consumed more steam.

DRAG—A line of (usually) heavy freight cars. The term was also applied to a slow, heavy freight train.

DRAWBAR—The stem, or shank, that holds the coupler in place. The term has been used interchangeably, though incorrectly, with "drawhead."

DRAWHEAD—A coupler; specifically the face, or non-moving portion, of the coupler.

DRIVERS—The large, powered wheels of a steam locomotive, located directly below the boiler. The drivers, or driving wheels, as they sometimes were called, carried most of the locomotive's weight and were fixed to one another on each side by a rod—the side rod. One set of drivers was additionally connected to the main rod (see).

DRIVER SHOES—The brake shoes that press against the driving wheels of an engine.

DWARF SWITCH—A low switch stand and lever used for changing the direction of a turnout.

EAGLE EYE—Slang for a locomotive engineer.

EIGHT WHEELER—Another name for a mogul engine (see).

EXPRESS CAR—Usually part of a passenger or mail train, the express car carried packages and, quite often, cash or gold. Several express companies had contracts with various railroads to ship such expedited freight, and the express cars themselves often carried an "express messenger"—an employee of the express company—to handle the forwarding and to look after the security of the shipments. In appearances, express cars looked much like baggage cars.

EXTRA (TRAIN)—A train that was not listed on the schedule, or a scheduled train that was more than twelve hours late.

EXTRA BOARD—A seniority list of crewmen who were otherwise unassigned to a regular run.

FACING POINT SWITCH—A turnout in which the diverging rails face the approaching train or car.

FIFTEEN BY TWENTY-TWO—The size, in inches, of the cylinder diameter and the piston stroke, respectively.

FIFTY PERCENT AIR—When the Railroad Safety Appliance Act was passed by Congress in 1893, it required, amongst other things, that railroad engineers be able to stop their trains from the cab. The law went into effect in 1900 and was interpreted as meaning that at least half of the cars of a freight train be equipped with air brakes, hence "fifty percent air."

FIRE-BOY—Slang for fireman, the engine crewman who stoked the boiler fire and kept watch on the boiler water level. See also "tallow bucket."

FIREBOX—A large, grated area under the rear of the boiler in which fuel was burned. The heat from the firebox transferred through the overhead crown sheet into the boiler water, thus creating steam.

FIRE DOOR—The heavy, steel door that opened into the firebox under the boiler head. The fire doors varied in size, but were always large enough to allow the fireman to throw in a scoop of coal.

FLAG BOTH WAYS—To place a flagman (see) in front of and behind a stopped train.

FLAGMAN—A crewmember whose job was to protect a stopped train. The flagman positioned himself several hundred yards in front of or in rear of his train with a red flag, lantern, and torpedoes, in order to warn any approaching train. Oftentimes the fireman served as the head-end flagman.

FLANGER—A wing-bladed device used to remove snow, or sometimes ballast, from between the rails.

FLIMSY—Common term for the tissue-thin paper on which train orders were written.

FLUE—Internal dry pipes that carried hot gasses from the firebox through the boiler. The greater number of flues meant a greater transfer of heat to the boiler water.

FROG—See "switch frog."

FULL PIT—A reference to the forward part of the coal bin on an engine tender.

FUSEE—A flare.

GANGWAY—The passageway between the cab and the tender.

GAY-CATS—Inexperienced hoboes.

GLIM—A railroad lantern.

GOAT—A small engine used for shuttling various pieces of equipment and cars in a yard area.

GUARD RAILS—A second set of track rails that ran parallel to the through rails. Guard rails normally were spiked down inside the running rails to catch an errant wheel that might have derailed. Guard rails most often were found on bridges, tight curves, and close clearances along the main line.

GUNNELS—Longitudinal support rods that formed part of the underframe of freight and passenger cars. Also known as rods.

HACK—Slang term for a caboose.

HELPER (ENGINE)—A second, or even third, engine assigned to help with a heavy train over a grade. Helpers could be coupled on at the front end, the rear end, or mid-train, depending on the circumstances of the grade, the curves, and the total tonnage. Helper engines in steam days each required their own engineer and fireman.

HIGHBALL—1.) To move with all possible speed; 2.) A signal to begin a run.

HOBO—1.) A nomadic traveler who rode trains surreptitiously without a ticket or without permission; 2.) A railroad laborer, usually a member of a traveling track gang.

HOG—A steam locomotive.

HOGGER—A locomotive engineer.

HOME (SIGNAL)—A trackside signal placed at the starting point of the territory that it controlled. In some cases, the home signal was augmented with a distant signal (see) that displayed the same indication.

HOOK—A long rake-like iron rod used by the fireman to arrange the bed of burning coal within the firebox.

HOOKING UP—The term referred to pulling back on the lever (see) in order to shorten the cut-off time of steam entering into the cylinders.

Hooking up usually took place on a downgrade or on a flat grade, thus taking advantage of momentum to move the train.

HOSTLER—A yard employee whose primary job was to move engines to and from the roundhouse, coal chute, water tank, or other maintenance areas.

HOT BOX—An overheated axle bearing.

HOUSE TRACK—The side track near a station that served a freight house or freight platform.

I—Telegraphic shorthand that essentially meant: "I acknowledge your call."

INJECTOR—A steam-operated device on the engine that forced water into the boiler under pressure.

INTERLOCKING (PLANT)—A system of interconnected switches and signals arranged in such a manner that made it impossible to set a signal to display "clear" when a switch was in the wrong position. The controlling interlocking levers, usually located in a tower, were attached to an array of notched bars and magnetic catches that required a specific operating sequence. Once a route had been set and signaled, no intervening track switch could be thrown to interfere with the intended route.

JACKET—The outer casing of the boiler.

JOHNSON BAR—See "lever."

JOURNAL BOXES—The bearing and lubrication enclosures at the ends of each axle.

JULL PLOW—A snow plow that utilized a diagonally-mounted auger conveyor.

KEY—The portion of the telegraph apparatus that the operator manipulated by tapping to produce an open and closed electrical circuit. This, in turn, created the staccato clicking of Morse code.

LEAD TRACK—A primary yard track that opened into several parallel tracks. Also known as a drill track, the switch engine used the lead as a back-and-forth track to sort cars into their proper groupings.

LEVER—Also known as the Johnson bar or reverse lever, it was mounted vertically in front of or alongside the engineer's seat. The lever controlled the cut-off point of steam entering into the engine cylinders on any given stroke. Adjusting the position of the lever was known as "notching her up" (pulling the lever back), or "shoving her in the corner" (setting the lever at the full forward, maximum power position). Pulling the lever to a rearward position on the quadrant would place the engine in reverse.

LIFT THE PIN—To pull the pin that locks the coupler in the closed position; to uncouple.

LIGHT ENGINE—An engine traveling without any cars coupled to it.

LIMITED—A generic name given to fast passenger/express/mail trains whose schedule called for a limited number of stops between the end terminals of the run. Limiteds were likely to whiz through small towns, picking up and dropping off a mail bag on the fly.

LINK AND PIN—An archaic coupling device that consisted of a heavy, elongated link and a locking pin. The link was placed in a coupling box located on the end of a car and the pin was dropped through a hole in the box to hold the link in place. It was a dangerous method of coupling and was outlawed after 1900 for cars in interchange service.

LOCAL—A freight train that picked up and set out cars at sidings and industries over a relatively short distance, then returned to the point of origin. The term can also be applied to a passenger train that ran a short distance out and back, stopping at all stations along the way.

LOOKING OVER THE AIR—Checking the air brake connections from car to car.

MAIN ROD—The heavy working beam that connected the piston rod to the driving wheels. It was the main rod that converted the back-and-forth motion of the piston rod to the rotary motion of the drivers (see "drivers"). The main rods were located on either side of the engine—one for each piston.

MARKERS—Signal lanterns hung on the rear of a train.

MOGUL—A steam locomotive with two pilot wheels on the front, followed by six larger driving wheels. Moguls were common freight engines during the turn-of-the-century era and saw occasional duty on slower passenger trains.

MOTOR—A motor car; a self-propelled baggage-mail car and coach, powered by external electricity or by a gasoline-generator combination.

ON THE CARPET—To stand before the superintendent to answer for a breach of procedure.

OPERATOR—The telegraph operator at a depot. The operator transmitted and copied both railroad business and telegrams. The operator was junior in seniority to the station agent, who usually worked the day shift. The operator often was required to perform routine bookkeeping and record keeping tasks at the station when not engaged in telegraphic chores.

OPERATOR'S TISSUE—See "flimsy."

ORDER BOARD—A signal device at a station or tower used to inform the crew of an approaching train that they had an order awaiting them.

ORDERS—Written instructions to the train crew, usually sent from the dispatcher by telegraph.

OS—Telegraphic shorthand used to report to the dispatcher that a train had arrived or passed a station. The station operator would send the OS followed by the station call sign, the train number, and the time. A typical call would read: "OS DE. No. 92 by at 9:15."

PALACE CAR—A first-class sleeping car, decorated in the ornate tastes of the late Victorian era. See "sleeper."

PILOT—The front platform of a locomotive; or the front set of guiding wheels under the same.

PIT—A trench for dumping coal or wood ash from the firebox. The ash pit was placed between the rails on a service track.

PONY TRUCK—The two small guide wheels located at the front of an engine under the pilot.

POUND THE BRASS—Operate a telegraph key.

POWER-YARD—The engine service and storage area of a railroad yard.

PULLED HIS DOOR—Opened the firebox door.

PULLING A DRAWBAR—Breaking a coupler shank, usually the result of too much strain or jerky operation.

RAKING THE FIRE—Smoothing out the burning coal bed in the engine firebox so as to insure an equal distribution of heat.

REVERSE LEVER—See "lever."

RODS—See 1.) "gunnels"; or 2.) "drivers."

ROTARY SNOWPLOW—A large, steam-powered device that utilized a revolving blade mounted on a horizontal axis to remove snow from the tracks and throw it to one side. The rotary was pushed by one or more engines.

ROUNDHOUSE—An enclosed, semi-circular service facility for steam locomotives. Engines entered their assigned stall in the roundhouse via a revolving turntable, which also served to face the engine in the proper direction.

RUN-BOARD, or RUNNING BOARD—A narrow walkway that ran along the top of boxcars, and along the side frames of certain other types of cars, thus enabling trainmen to traverse the length of a moving train. Running boards were not installed on flatcars or gondolas, as the trainmen could climb over the car's cargo.

SAFETY VALVE—A device located on top of the locomotive that would release excess steam from the boiler. The release pressure could be adjusted by a tensioning spring. Also known as a "pop" or "relief valve."

SAND-BOX—A protruding bin on top of the engine boiler that held sand. The sand was piped down to the rail, in front of the drivers. The engineer could open the sand valve and thus dust the rails, which would provide better adhesion for the drivers.

SAWING (or SAWING BY)—A back and forth switching maneuver that was used when two trains were to pass one another, but when one or both trains were longer than the available passing siding.

SECTION (OF A TRAIN)—If the total tonnage of a train movement made it necessary to dispatch two or more trains on the same schedule (with a safety interval of several minutes between them), then each train would be dispatched as first, second, third, etc., section of a given train number. In the case of, say, train number 25, with two sections, those two trains would be called first No. 25 and second No. 25, and written as 1/25 and 2/25.

SECTION (OF TRACK)—A territory of perhaps ten or fifteen miles that was maintained by a track gang. The gang worked under the authority of a section foreman. There usually was a section house located approximately midway along the section where the track gang ate and slept.

SET-DOWN—An invitation to a hobo to sit down at a table for a meal. A set-down usually was offered by a sympathetic family or individual.

SET HIS AIR—An engineer applying the air brakes to the train.

SHOW RED—To display a red board, semaphore signal, or lantern at a station; the signal for an approaching train to stop. Showing red was the opposite of a white (later, green) signal, which indicated a clear track ahead.

SHUT OFF—Close the throttle, thereby slowing or stopping the train.

SIDE ROD—See "drivers."

SIDING—A side track. Some sidings were double-ended (i.e., attached by a switch to the main track at either end) and commonly were used for passing purposes. A stub siding diverted from the main track only at one end and often served a trackside industry.

SLACK—The inherent looseness of couplings, which increases dramatically with each additional car. Slack can "run in" when the train slows, or "run out" when the train accelerates.

SLEEPER, or SLEEPING CAR—A passenger car that could be converted from seats in the daytime to beds at night. The berths usually were set in two tiers, bunk style, with a partition between sections and heavy privacy drapes facing the aisle. Some sleeping cars featured staterooms or drawing rooms—semi-permanent spaces with bulkheads and doors.

SLOW ORDER—A directive from the dispatcher that all trains were to run at a reduced speed through a certain stretch of track. Slow orders usually were issued due to track maintenance, localized flooding, or other anomalies. The term, "put the orders on him," referred to an engineer thus directed to run slow.

SNOW SHED—A timbered roof built over the tracks to prevent the accumulation of snow. Snow sheds were common along mountain cuts.

SOUNDER—A component of a telegraph set, the sounder was the electromechanical device that emitted the clicking cadence sound of Morse code. The sounder usually was mounted on a small L-shaped wooden shelf that, in turn, was attached to a scissors-type extension arm. It was traditional, even necessary, to amplify the clicking sound by wedging a tin tobacco can between the sounder and back of the wooden shelf.

SPANNER—A wrench.

STATION REPORT—A routine report that listed, among other things, the status of cars set out or picked up.

STAY BOLT—A long-shanked spacer bolt that held the inner boiler to the outer shell.

SWITCH—A turnout; a geometric arrangement of track utilizing the lateral movement of rails in such a way that it allowed a train or car to divert (switch) from one track to an adjoining track. "Throwing a switch" meant aligning the direction of a turnout.

SWITCHBACK—A track arrangement utilized on very steep grades. A switchback was essentially a zig-zag configuration, whereby a train must move forward on one leg, then back through the next leg, all the while climbing or descending the grade.

SWITCHER—A locomotive designed for shuffling cars in a yard. Switchers were easy to spot because they did not have small pilot wheels on the front, as did road engines.

SWITCH FROG—The V-shaped portion of a turnout where the two inside rails diverge from one another.

'TABLE—See "turntable."

TAKE SIDING—Common term for a train to switch off the main track onto a siding and await the passage of another train.

TAKING WATER—Filling the water tank on an engine tender.

TALLOW BUCKET—A large can for carrying rendered animal fat. The tallow was used for lubricating moving parts on the engine. The tallow was kept warm and fluid by storing it on a shelf on the boiler backhead. Since the fireman sometimes was required to step out onto the moving engine and lubricate the rods, the name "tallow-pot" was used as yet another slang term for that position.

TAMP BALLAST—Shoring up and leveling track by forcing ballast under the ties, usually with a shovel and a pry rod.

TARGET—A colored blade, often with a lantern attached. The target was attached to a switch stand, and revolved to display a different color according to the direction that the turnout was set.

TELESCOPED—The tendency of railroad cars in a collision to crush through one another, overlapping like the sections of a telescope.

TENDER—The trailing segment of the engine. The tender was an integral part of a steam engine, and was detached only when in the shop for repairs or maintenance. The tender consisted of two bunkers—one for fuel and the other for water. Also called the "tank."

TEN-WHEELER—A steam locomotive with four small pilot wheels on the front, followed by six larger driving wheels. Ten-Wheelers generally were used in passenger service or on fast freights in the late nineteenth century.

THROTTLE—The lever used to control the amount of steam passing from the boiler to the cylinders. The throttle was mounted horizontally on the boiler backhead, and was adjusted by the engineer's left hand.

TIME FREIGHT—A freight train that carried a perishable or important commodity. A time freight was given priority over other freights in order to adhere to a close schedule, much the same as a passenger train. Time freights usually made only a few scheduled stops between division points, most often to take on water or coal.

TISSUE—See "flimsy."

TORPEDO—A small, noise emitting explosive device that was strapped to the railhead. The torpedo exploded with a loud report when run over, thus warning the engine crew of a danger ahead.

TRACKWALKER—A track inspector who carried tools to make on-the-spot repairs to such things as loose bolts and spikes.

TRAIN BUTCHER—A boy who sold newspapers, magazines, dime novels, and candy aboard a passenger train.

TRAIN ORDERS—The written instructions to a train crew, telling them such things as where to take a siding, where to meet another train, and when to leave or arrive at a given point. Generally, train orders covered contingencies that were not specified in a published timetable. The most common train order, the Form 19, was written on thin, onion-skin paper and passed to the engineer and conductor via a small, wooden hoop that was snagged by the crewmen. Hence the term, "hooping up orders."

TRAIN ORDER STATION—A station designated to copy orders from the dispatcher and convey them to a passing train. Not every station was a train order station.

TRAIN (REGISTER) SHEET—A ledger list of all the trains running in a given territory over time. The dispatcher used the sheet to keep tabs on the trains in his division, as did some agent-operators.

TRIPLE VALVE—An air brake pressure regulating device located under each car.

TRUCK—The wheel and suspension assembly under each end of a car. Most freight cars had four-wheel, two-axle trucks, while some heavyweight passenger cars sported six-wheel, three-axle trucks.

TURNTABLE—Sometimes referred to as the "table," the turntable was a revolving platform located in the center of a roundhouse facility.

The turntable moved in a circular motion to line up with tracks running into the several engine stalls of the roundhouse. With an engine sitting upon it, the turntable could revolve around and completely reverse the direction that the engine faced. It could also line up with tracks that led away from the roundhouse and into the adjoining yard area, thus pointing an engine to an outbound track.

UP IN THE AIR—1.) Slang for moving fast; 2.) Extremely excited.

VAN—See "caboose."

VAPOR CLOCK—One of several slang terms for the steam pressure gauge. At the time of our stories, 200 pounds per square inch of steam pressure was typical of most large engines.

WASH OUT SIGNAL—A lantern or hand signal calling for an emergency stop.

WEDGE PLOW—A large point-bladed snowplow that was pushed ahead of an engine.

WESTINGHOUSE—George Westinghouse, Jr. (1846-1914), the inventor of the railroad airbrake.

WILDCAT (ENGINE)—See "light engine."

WIPER—A roundhouse employee who wiped down an engine after its run.

Y—A three-pronged track arrangement that allowed an engine or short train to back down one leg and pull forward on the other, thus reversing its direction.

YARD LIMIT—An area in which certain mainline running rules did not apply. For instance, inside a yard limit area it was not necessary to send out a flagman to protect the front and rear of a train when the train was stopped.

YEGG—A hobo thief or criminal.

About the Author

As a boy, author Michael Gillespie was mesmerized by the steam trains that daily crested the hill behind his home near Independence, Missouri. He never got over that fascination.

While in high school, Mike "volunteered" to assist the agent at the local depot and while in college he worked for Amtrak in Kansas City's massive Union Station. He would later write about both adventures in articles for *Classic Trains* magazine.

A high school history teacher-turned-writer, Mike began collecting long-forgotten stories of railroad lore as a means of learning more about the history of the railways.

As his wife and kids well know, a Sunday drive with Mike almost always leads to the banks of the Missouri River or to a roadway that runs parallel to a railroad.

Publisher's note:

Michael's interest in railroad history coincides with his studies of river steamboat history. He has written two books on that subject and served for a year as the onboard historian for the steamboat *American Queen*.

Mike Gillespie's first book, *Wild River, Wooden Boats,* has been described as "one of the very best steamboating histories available for the Missouri River." His second book is the highly popular *Come Hell or High Water,* a lively history of steamboating on the Mississippi River and Ohio River.

Michael, a witty, talented, and knowledgeable historian and storyteller, passed away in 2012. He was determined, that year, to finish compiling and editing this collection of Old Time Railroad Stories from the era of steam, which we are publishing posthumously in three parts in print and as a digital collection. Each of his anthologies provides entertaining reading and copious collections of historic photographs, charts and sketches.

Michael's first book, *Old Time Railroad Stories,* has been praised by rail lovers around the country. We are proud to now release his second volume of stories as *The Phantom Brakeman*. The third and final book in the collection, *Dead Under His Cab,* is currently in production.

All of Michael's books can be purchased online
at www.greatriverarts.com > SHOP

Real Photo Note Cards

See online at www.greatriverarts.com

5" x 7" real photo note cards in a variety of Birding, Heritage and River Travel subjects. Individually packaged in clear acetate sleeves. See at www.greatriverarts.com. . . . . . . $4.60 each

Discover! America's Great River Road

by Pat Middleton

The classic guides to the heritage and natural history of the Mississippi River Valley and the scenic GREAT RIVER ROAD. Approximately 240 pages in each volume . . . . $19.95 each

Vol 1 ~ The Upper Mississippi River

Vol 2 ~ The Middle Mississippi River

Vol 3 ~ The Lower Mississippi River

Vol 4 ~ Arkansas, Mississippi and Louisiana

The Mississippi River Activity Guide for Kids

With teacher notes . . . . $19.95 each

by Pat Middleton

Watch the river come alive for your children or students as they

- learn how locks and dams work;
- find their way through a maze of river sloughs;
- create a river map;
- identify leaves, plants, animals, and fish along the river.